Improvements in Bio-Based Building Blocks Production Through Process Intensification and Sustainability Concepts

Improvements in Bio-Based Building Blocks Production Through Process Intensification and Sustainability Concepts

JUAN GABRIEL SEGOVIA-HERNÁNDEZ
Department of Chemical Engineering, University of Guanajuato, Guanajuato, Mexico

EDUARDO SÁNCHEZ-RAMÍREZ
Department of Chemical Engineering, University of Guanajuato, Guanajuato, Mexico

CÉSAR RAMÍREZ-MÁRQUEZ
Department of Chemical Engineering, University of Guanajuato, Guanajuato, Mexico

GABRIEL CONTRERAS-ZARAZÚA
Department of Chemical Engineering, University of Guanajuato, Guanajuato, Mexico

ELSEVIER

Elsevier
Radarweg 29, PO Box 211, 1000 AE Amsterdam, Netherlands
The Boulevard, Langford Lane, Kidlington, Oxford OX5 1GB, United Kingdom
50 Hampshire Street, 5th Floor, Cambridge, MA 02139, United States

British Library Cataloguing-in-Publication Data
A catalogue record for this book is available from the British Library

Library of Congress Cataloging-in-Publication Data
A catalog record for this book is available from the Library of Congress

ISBN: 978-0-323-89870-6

For Information on all Elsevier publications
visit our website at https://www.elsevier.com/books-and-journals

Publisher: Susan Dennis
Acquisitions Editor: Anita Koch
Editorial Project Manager: Andrea R. Dulberger
Production Project Manager: Bharatwaj Varatharajan
Cover Designer: Christian J. Bilbow

Typeset by MPS Limited, Chennai, India

Contents

Author biographies

Juan Gabriel Segovia-Hernández, Professor in the Department of Chemical Engineering of University of Guanajuato (México), has expertise in the synthesis, design, and optimization of bioprocesses. He has contributed to defining systematic methodologies to found, in a complete way, optimum sustainable and green processes for the production of several commodities. He also applied his methodologies to the production of biofuels and biobased building blocks. Products of his research are more than 140 papers published in high-impact-factor indexed journals, 3 books with prestigious international publishers, and 4 patent registers. In addition, he acts as a reviewer for more than 25 top journals in chemical engineering, energy, and applied chemistry. For his pioneering work and remarkable achievements in his area of scientific research, he was elected National President of the Mexican Academy of Chemical Engineering (2013–15) and has been a member of the Mexican Academy of Sciences since 2012. He also is the Associate Editor of *Chemical Engineering and Processing: Process Intensification Journal and Chemical Engineering Research and Design Journal* (Elsevier).

Eduardo Sánchez-Ramírez, Professor at the Department of Chemical Engineering at the University of Guanajuato (Mexico) since 2017, has gained considerable experience in the area of synthesis, design, simulation, control, and optimization of chemical processes. Currently published contributions focus on the production of biofuels and base chemicals in the chemical industry. He has currently published more than 35 articles in indexed journals, 9 book chapters with renowned publishers, and has registered 2 patents. He acts as a reviewer of indexed journals in the area of energy and chemical engineering. He earned his PhD summa cum laude in 2017 and his National System of Researchers Level 1, SNI Mexico in 2019. He was Guest Editor of *Chemical Engineering and Processing Process Intensification Journal* in May 2020.

César Ramírez-Márquez, Research Associate in the Chemical Engineering Department of the University of Guanajuato, Mexico, holds an MEng and a PhD from the University of Guanajuato, Mexico, under the supervision of Prof. J.G. Segovia-Hernández. His research focuses on synthesis, design, simulation, optimization, and control of chemical processes. Currently published contributions focus on the production of materials for the solar energy industry and base chemicals in the chemical industry. He has currently published more than 25 journal papers, 4 book chapters and presented work at more than 10 international/regional conferences, and registered four patents.

Gabriel Contreras-Zarazúa, Professor in the Department of Chemical Engineering at the University of Guanajuato, Mexico, since 2017, obtained his PhD degree at the University of Guanajuato, Mexico. He obtained his PhD degree summa cum laude in 2020. His research is focused on design, simulation, and optimization of biorefineries to produce high-value products. Additionally, he has experience in control analysis, life cycle and safety assessments, and design and optimization of supply chains applied to bioprocesses. Currently, he has published more than 15 articles in indexed journals, 2 book chapters, and has registered 1 patent. He is a reviewer of indexed journals in the areas of chemical engineering and process intensification.

CHAPTER 1

Why are bio-based chemical building blocks needed?

Contents

1.1 Are bio-based chemical building blocks needed?

Continuous, inspiring, and interconnected step-by-step changes in thought and understanding, know-how, actions, and behavior have often been instrumental in transitions from one particular age to the next in human history. This also applies to the present century and its sustainability challenges at the planetary, regional, and local levels. Therefore, it is of great importance and relevance to move forward on the journey that has been started globally to address a not insignificant number of challenges. It is, however, essential to go beyond descriptive work by continuing with novel, inspiring, and interconnected steps to find solutions to overcome these challenges. As this huge task also requires multidimensional communication, understanding, and actions across different regions, cultures, disciplines, and knowledge areas, the development of a common conceptual framework such as the concept of bioeconomy has been accepted globally as very valuable (Sierra et al., 2021).

The general orientation provided by the United Nations Sustainable Development Goals (SDGs) is of special interest and a call to action for all stakeholders to reach the SDGs by 2030. Throughout history, finding

Improvements in Bio-Based Building Blocks Production Through Process Intensification and Sustainability Concepts
DOI: https://doi.org/10.1016/B978-0-323-89870-6.00003-2

innovative solutions based on available biological resources to successfully overcome challenges has been a hallmark of human achievements. Therefore, the accumulated traditional knowledge and skills and the rapid advances in the life sciences have made biotechnology a key enabling technology toward improving the quality of life. Bioeconomy, bringing together bioresources, biotechnology, ecosystems, and economy, has emerged as an attractive top-level political concept for creating, developing, and revitalizing economic systems worldwide by making use of renewable biological resources in a sustainable way. As there is no universal definition, we are very much in line with the one adopted by the Global Bioeconomy Summit in 2015, which defined bioeconomy as "the knowledge-based production and utilization of biological resources, innovative biological processes and principles to sustainably provide goods and services across all economic sectors" (Wohlgemuth et al., 2021).

The evolution from political objectives of a bioeconomy that is based on knowledge in the relevant sciences and technologies, industries, and societies to bioeconomy policies, strategies, and initiatives has been spreading rapidly worldwide. Biotechnology is a key enabling technology, not only for highly developed and diversified bioeconomies but also for advanced and basic primary sector bioeconomies. From the given definition, it is obvious that bioeconomy is much more than biotechnology and includes other sciences, but it also goes beyond innovations in sciences and technologies by incorporating industrial, organizational, political, and social innovations.

An important aspect of bioeconomy is not only to collect and summarize existing knowledge but also to present and evaluate new strategies and technological processes, and to suggest and to select the best directions of change and development, taking into account global, regional, and local specificities. Resource distribution and allocation, increasing specialization, and facilitated transportation and trade worldwide have shown the strengths of exchanging and sharing products and provided tremendous opportunities for economic growth. As well as strengths, these developments have also demonstrated imbalances, weaknesses, and threats to global supply chains. A holistic and innovative bioeconomy approach takes into account various perspectives across different disciplines of science, industry, and society to make science-based political decisions and create sustainable new opportunities and value creation chains. Large economic value is created from biotransformations of bio-based and fossil-based resources to intermediaries and final products, which enter the industrial manufacturing chain towardas suitable goods. This value creation arises in many areas such as health, nutrition, materials, energy, and environment in ever more complex and diversified bioeconomy networks. Bio-based industries is a key enabling technology in all of these bioeconomy networks (Sierra et al., 2021).

Bioprocesses arose from the field of zymotechnology, which began as a search for a better understanding of industrial fermentation, particularly in relation to the brewing of beer. Beer was an important industrial, and not just social, commodity. In late-19th century Germany, brewing contributed as much to the gross national product as did steel, and taxes on alcohol proved to be significant sources of revenue. In the 1860s, institutes and remunerative consultancies were dedicated to the technology of brewing. The most famous was the private Carlsberg Institute, founded in 1875, which employed Emil Christian Hansen, who pioneered the pure yeast process for the reliable production of consistent beer. Less well known were private consultancies that advised the brewing industry. One of these, the Zymotechnic Institute, was established in Chicago by the German-born chemist John Ewald Siebel.

The expansion of zymotechnology continued during World War I in response to industrial needs to support the war. Max Delbrück grew yeast on an immense scale during the war to meet 60% of Germany's animal feed needs. Compounds of another fermentation product, lactic acid, made up for a lack of hydraulic fluid, glycerol. On the Allied side, the Russian chemist Chaim Weizmann used starch to eliminate Britain's shortage of acetone, a key raw material for cordite, by fermenting maize to acetone. The industrial potential of fermentation outgrew its traditional home in brewing, and zymotechnology soon gave way to "biotechnology" (Doran, 2012).

With food shortages spreading and resources fading, some dreamed of a new industrial solution. The Hungarian Károly Ereky coined the word "biotechnology" in Hungary in 1919 to describe a technology based on converting raw materials into a more useful product. He built a slaughterhouse for 1000 pigs and also a fattening farm with space for 50,000 pigs, and he raised over 100,000 pigs a year. The enterprise was enormous, becoming one of the largest and most profitable meat and fat operations in the world. In his book, *Biotechnologie*, Ereky further developed a theme that would be reiterated through the 20th century: biotechnology could provide solutions to societal crises, such as food and energy shortages. For Ereky, the term "biotechnology" indicated the process by which raw materials could be biologically upgraded into socially useful products.

This catchword spread quickly after World War I, as the word "biotechnology" entered German dictionaries and was taken up abroad by business-hungry private consultancies as far away as the United States. In Chicago, for example, the coming of prohibition at the end of World War I encouraged biological industries to create opportunities for new fermentation products, in particular a market for nonalcoholic drinks. Emil Siebel, the son of the founder of the Zymotechnic Institute, broke away from his father's company to establish the "Bureau of Biotechnology," which specifically offered expertise in fermented nonalcoholic drinks.

The belief that the needs of an industrial society could be met by fermenting agricultural waste was an important ingredient of the "chemurgic movement." Fermentation-based processes generated products of ever-growing utility. However, this flourishing industry of fermentation to obtain bioproducts began to compete economically with the advantages of obtaining the same products from nonrenewable sources (oil) in a more efficient way and at a lower cost (Doran, 2012).

In recent years, due to the large increase in petroleum cost, there has been a reemergence of interest in large-volume production of fermentation chemicals. Biotechnology is providing new, low-cost, and highly efficient fermentation processes for the production of chemicals from biomass resources. Moreover, with a wide range of microorganisms already available and much more recently discovered, the fermentation of sugars represents an important route for the production of new bioproducts. However, the current economic impact of fermentation bioproducts is still limited, in large part a result of difficulties in product recovery. Thus, substantial improvements to existing recovery technology are needed in order to allow chemicals from fermentation to penetrate further in the organic chemical industry (Corma et al., 2007).

The bio-based industry is an emerging sector organized around interconnected value chains, which aims to transform renewable biological feedstock, such as forestry, agricultural, and aquatic biomass, as well as sidestreams and byproducts from industrial bioprocessing, and other residues such as sludge and municipal waste, into bio-based products, materials, fuels, and energy, replacing their fossil-based counterparts. They offer a huge potential to tackle societal and environmental challenges and, additionally, play an important role in stimulating sustainable growth and boosting the competitiveness of countries by reindustrializing and revitalizing rural and coastal areas and providing new job opportunities.

The reality that can be seen in these bio-based projects is that new value chains are much more interconnected. These value chains arise from the connections between different types of feedstock and different processing and biorefining technologies, transforming them into a wide variety of bio-based chemical building blocks (CBBs), materials, food and feed ingredients, and consumer products (e.g., cosmetics) for a wide range of market sectors, thereby producing an ever-increasing number of new bio-based value chains. This also corresponds to the reality of the bio-based sector development; hence the relevance of CBBs in the concept of bioeconomy.

A CBB is a molecule that can be converted to various secondary chemicals and intermediates, and, in turn, into a broad range of different downstream uses. The largest markets for bio-based CBBs are in the production of bio-based polymers, lubricants, and solvents. This chapter looks at two types of bio-based CBBs: drop-in bio-based chemicals and novel bio-based chemicals (United States Department of Energy Energy Efficiency and Renewable Energy, 2004).

1.1.1 Drop-in bio-based chemicals

Drop-in chemicals are bio-based versions of existing petrochemicals that have established markets. As they are chemically identical to existing hydrocarbon-based products, their use can reduce financial and technological risks and promote faster access to markets for producers.

1.1.2 Novel bio-based chemicals

Novel bio-based chemicals bear higher financial and technological risks for producers but can be used to produce products such as aconic acid and methylenesuccinic acid that cannot be obtained through traditional chemical reactions and products that may offer unique and superior properties that are unattainable with fossil-based alternatives, such as biodegradability.

There is an existing market for CBBs, but it can be considered relatively immature, with development levels varying according to the building block considered and ranging from proof of concept in the laboratory to full commercial production. Strong cooperation within the value chain from feedstock producer to end user is required for new CBBs to successfully enter the market.

In 2013, the demand for bio-based CBBs in Europe was 1029 MEUR, equivalent to 35% of the total global production. The market grew at a compound annual growth rate of approximately 18.6% per annum between 2008 and 2015. It has been estimated that by 2030 the bio-based CBBs market in Europe could reach between 4.8 and 10.4 BEUR. The market value could be greater than this if the various hurdles to the development of bio-based CBBs are addressed. The greatest driver for the market uptake of bio-based CBBs is to overcome increasing volatility in fossil-fuel price and supply. Market prices for chemicals rise when fossil supply is tight, so the subsequent increasing uncertainty and volatility of crude oil prices is likely to push commodity chemical companies toward bringing in alternatives to traditional fossil fuels to ensure that their customers have a stable product supply.

One of the key hurdles to the production of bio-based CBBs is that of feedstock availability and cost. Most of the currently available bio-based CBBs are based on commodity agricultural products such as sugars and vegetable oils which can vary significantly in price and are expensive. Given that many CBBs are bulk chemicals, a large amount of feedstock will be needed. There are concerns by some that the potential for supplying extra sugar and oils is limited, though others believe that there is still much potential for yield improvement in such commodities. The use of waste and residue streams would be attractive as they are both cheap and widely available. The ability to interchange feedstocks according to availability would also be useful. Many technical challenges, especially relating to downstream processing need to be overcome to help promote the use of alternative feedstock streams and reduce processing costs, but even if these challenges are successfully addressed, it will be necessary to persuade highly conservative processors to change the production process to accommodate a new feedstock or a product with new properties. A combination of high feedstock, conversion and downstream processing costs mean that the cost of producing bio-based chemicals is currently more expensive than processes using fossil-fuel feedstocks. Opportunities for bio-based premiums to overcome price differentials for CBBs are considered to be lower than for other markets, for instance bioplastics, because the CBB producer is further away from the final consumer.

The greatest hurdle to promoting investments in bio-based chemicals however is regulatory uncertainty and instability through its effects on pricing and demand for products. Vagueness in terms such "waste", "residues", and "green" fail to provide sufficient investment certainty, while at present incentives for bioenergy and biofuel markets mean that these sectors can pay more for feedstock than nonincentived markets such as bio-based chemicals resulting from feedstock prices that are artificially inflated.

Current global CBB (excluding biofuels) is estimated to be around 50 million tons. Notable examples of bio-based chemicals include non-food starch, cellulose fibers, and cellulose derivatives, tall oils, fatty acids, and fermentation products such as ethanol and citric acid. However, the majority of organic chemicals and polymers are still derived from fossil-based feedstocks, predominantly oil and gas. Nonenergy applications account for around 9% of all fossil-fuel (oil, gas, and coal) use and 16% of oil products. Global petrochemical production of chemicals and polymers is estimated at around 330 million tons. The primary output is dominated by a small number of key building blocks, namely methanol, ethylene, propylene, butadiene, benzene, toluene, and xylene (United States Department of Energy Energy Efficiency and Renewable Energy, 2004). These building blocks are mainly converted to polymers and plastics but they are also converted to a staggering number of different fine and specialty chemicals with specific functions and attributes. From a technical point of view, almost all industrial materials made from fossil resources could be substituted by their bio-based counterparts. However, the cost of bio-based production in many cases exceeds the cost of petrochemical production. Also, new products must be proven to perform at least as good as the petrochemical equivalent they are substituting and to have a lower environmental impact.

Historically bio-based chemical producers have targeted high-value fine or specialty chemicals markets, often where specific functionality played an important role. As proof of the aforementioned Fig. 1.1 shows some of the most common sugar obtained from biomass and their respective downstream processes used to produce high-value commodities. Currently, the low price of crude oil acted as a barrier to bio-based commodity chemical production and producers focused on the specific attributes of bio-based chemicals such as their complex structure to justify production costs.

Bio-based CBB can be classified based on a number of their key characteristics (International Energy Agency – Bioenergy Report, 2007). Major feedstocks include perennial grasses, starch crops (e.g., wheat and maize), sugar crops (e.g., beet and cane), lignocellulosic crops (e.g.,

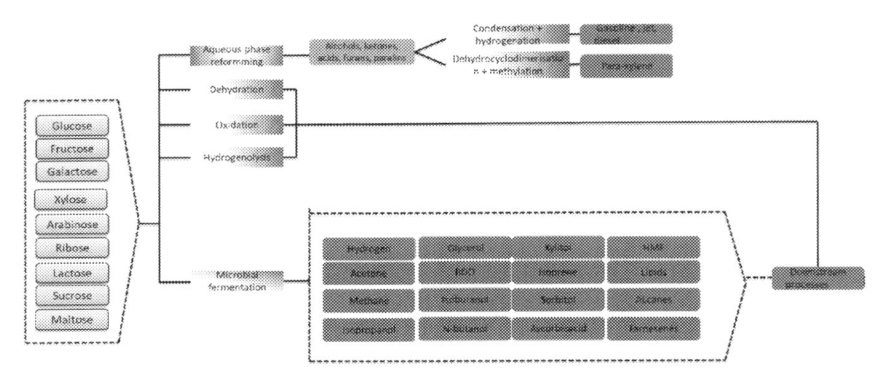

Figure 1.1 Main sugars obtained from biomass and their respective main downstream processes.

managed forest, short rotation coppice, switch grass), lignocellulosic residues (e.g., stover and straw), oil crops (e.g., palm and oilseed rape), aquatic biomass (e.g., algae and seaweeds), and organic residues (e.g., industrial, commercial, and post-consumer waste). These feedstocks include five- and six-carbon carbohydrates from starch, sucrose, or cellulose; a mixed five- and six-carbon carbohydrates stream derived from hemicelluloses, lignin, oils (plant-based or algal), organic solutions from grasses, pyrolytic liquids.

1.1.3 C6 and C6/C5 Sugar

Six-carbon sugar platforms can be accessed from sucrose or through the hydrolysis of starch or cellulose to give glucose. Glucose serves as feedstock for (biological) fermentation processes providing access to a variety of important CBBs. Glucose can also be converted by chemical processing to useful CBBs. Mixed six- and five-carbon platforms are produced from the hydrolysis of hemicelluloses. The fermentation of these carbohydrate streams can in theory produce the same products as six-carbon sugar streams; however, technical, biological, and economic barriers need to be overcome before these opportunities can be exploited. Chemical manipulation of these streams can provide a range of useful molecules.

1.1.3.1 Fermentation products

The number of CBBs accessible through fermentation is considerable. Fermentation has been used extensively by the chemical industry to produce several products with chemical production through fermentation. Modern biotechnology is allowing the industry to target new and previously abandoned fermentation products and improve the economics of products with commercial potential. Coupled with increasing fossil feedstock costs, cost reductions in the production of traditional fermentation products, such as ethanol and lactic acid, will allow derivative products to capture new or increased market shares. Improving cost structures will also allow previously abandoned products such as butanol to reenter the market. Many see the future abundant availability of carbohydrates derived from lignocellulosic biomass as the main driver. Fermentation also gives the industry access to new CBBs previously inaccessible due to cost constraints. The development of cost-effective fermentation processes to succinic, itaconic, and glutamic acids promises the potential for novel chemical development.

1.1.3.2 Chemical transformation products

Six- and five-carbon carbohydrates can undergo selective dehydration, hydrogenation, and oxidation reactions to give useful products, such as: sorbitol, furfural, glucaric acid, hydroxymethylfurfural, and levulinic acid (International Energy Agency – Bioenergy Report, 2007).

1.1.4 Plant-based oil

The oleochemical industry is a major producer of bio-based products. The majority of fatty acid derivatives are used as surface active agents in soaps, detergents, and personal care products. Major sources for these applications are coconut, palm and palm kernel oil, which are rich in C12–C18 saturated and monounsaturated fatty acids. Important products of unsaturated oils, such as: soybean, sunflower and linseed oil, include alkyd resins, linoleum, and epoxidized oils. Rapeseed oil, high in oleic acid, is a favored source for biolubricants. Commercialized bifunctional building blocks for bio-based plastics include sebacic acid and 11-aminoundecanoic acid, both from castor oil, and azelaic acid derived from oleic acid. Dimerized fatty acids are primarily used for polyamide resins and polyamide hot melt adhesives. In applications such as lubricants and hydraulic fluids, plant oil can act as a direct replacement for mineral (petroleum-derived) oil or require only minor chemical modification. As a chemical feedstock, the triacylglycerol molecule—the major component of most plant oils—is either (1) cleaved to glycerol and fatty acids or (2) converted to alkyl esters and glycerol by transesterification. The utility of the fatty acids and esters is determined primarily by their chain length and functionality. Given advances in plant genetics and oil processing, there is considerable interest in developing plant oils for the manufacture of polymers such as polyurethanes, polyamides, and epoxy resins. There is also an important subcategory of oilseeds that produce natural waxes, such as liquid wax from jojoba seeds and solid waxes collected from the leaf surfaces of the Carnuba Palm and several desert shrubs. These tend to be used in specialized high-value applications, such as cosmetics. Their excellent lubricity and stability in lubricant applications have led to interest in engineering wax ester production in commercial oilseed crops (International Energy Agency – Bioenergy Report, 2007).

1.1.5 Algae oil

There are more than 40,000 different algae species both in seawater and freshwater. Algae biomass can be a sustainable renewable resource for chemicals and energy. The major advantages of using microalgae as a renewable resource are:

1. compared to plants, algae have higher productivity;
2. microalgae can be cultivated in seawater or brackish water on nonarable land; and do not compete for resources with conventional agriculture;
3. the essential elements for growth are sunlight, water, CO_2, and inorganic nutrients such as nitrogen and phosphorous which can be found in residual streams;
4. the biomass can be harvested during all seasons and is homogenous and free of lignocellulose.

The main components of microalgae are species dependent but can contain a high protein content, quantities can be upto 50% of dry weight in growing cultures with all 20 amino acids present. Carbohydrates as storage products are also present and some species are rich in storage and functional lipids, they can accumulate upto 50% lipids, and in very specific cases upto 80% (the green algae *Botryococcus*) accumulates long chain hydrocarbons. Other valuable compounds include pigments, antioxidants, fatty acids, vitamins, antifungals, antimicrobials, antiviral toxins, and sterols (International Energy Agency – Bioenergy Report, 2007).

1.1.6 Organic solutions

The organic solution (press juice) contains valuable compounds, such as carbohydrates, proteins, free amino acids, organic acids, minerals, hormones, and enzymes depending on whether the biomass used as the feedstock is fresh or silage. Soluble carbohydrates and proteins—main components of fresh plant juice—can be used as a fermentation medium or for generating feed products. Silage press juice has been demonstrated as feedstock for the production of biochemicals and fuels. The organic solution (press juice) contains valuable compounds, such as carbohydrates, proteins, free amino acids, organic acids, minerals, hormones, and enzymes depending on whether the biomass used as the feedstock is fresh or silage. Soluble carbohydrates and proteins—main components of fresh plant juice—can be used as fermentation medium or for generating feed products. Silage press juice has been demonstrated as feedstock for the production of biochemicals and fuels. Lactic acid and its derivatives, as well as proteins, amino acids, bioethanols, and energy via anaerobic digestion, are the most favorable end products from the organic solution platform (International Energy Agency – Bioenergy Report, 2007).

1.1.7 Lignin

Lignin offers a significant opportunity for enhancing the operation of a lignocellulosic biorefinery. It is an extremely abundant raw material contributing as much as 30% of the weight and 40% of the energy content of lignocellulosic biomass. Lignin's native structure suggests that it could play a central role as a new chemical feedstock, particularly in the formation of supramolecular materials and aromatic chemicals. Upto now, the vast majority of industrial applications have been developed for lignosulfonates. These sulfonates are isolated from acid sulfite pulping and are used in a wide range of lower value applications where the form but not the quality is important. Around 67.5% of world consumption of lignosulfonates is for dispersant applications followed by binder and adhesive applications at 32.5%. Major end-use markets include construction, mining, animal feeds, and agriculture uses. The use of lignin for chemical production has so far been limited due to contamination from salts, carbohydrates, particulates, volatiles, and the molecular weight distribution of lignosulfonates. The only industrial exception is the limited production of vanillin from lignosulfonates. In addition to lignosulfonates, kraft lignin is produced as a commercial product for use as an external energy source and for the production of value-added applications. The production of more value-added chemicals from lignin (e.g., resins, composites and polymers, aromatic compounds, and carbon fibers) is viewed as a medium- to long-term opportunity that depends on the quality and functionality of the lignin that can be obtained (International Energy Agency – Bioenergy Report, 2007).

1.1.8 Pyrolysis oil

Biomass pyrolysis is the thermal depolymerization of biomass at modest temperatures in the absence of added oxygen. The spectrum of products from biomass pyrolysis depends on the process temperature, pressure, and residence time of the liberated pyrolysis vapors. A biorefinery based on pyrolysis oil is designed much like a traditional refinery. First biomass is converted into pyrolysis oil, which can be a decentralized process. Second, pyrolysis oil from different installations is collected at the biorefinery where it will be divided into different fractions. Each fraction can be upgraded with a different technology to finally derive the optimal combination of high-value and low value products from the pyrolysis oil. The major high-value compounds which are foreseen are phenols, organic acids, furfural, Hidroximetilfurfural (HMF), and levoglucosan. The major advantages of apyrolysis transformation is the possibility of decentralized production of the oil in regions where abundant biomass is readily available, making it possible to keep the minerals in the country of origin and creating the possibility of cost-effective transport of the resulting liquids. The basis for creating high-value compounds is the cost-effective fractionation of the pyrolysis oil. Fractionation will result in various qualities of oil needed for further upgrading into fine chemicals, petrochemicals, automotive fuels and energy (International Energy Agency – Bioenergy Report, 2007).

In Fig. 1.2, the general classification of bio-based CBBs is shown in summary form.

In summary, although many chemicals that are currently produced from petroleum will likely be produced from biomass in the future, several challenges still remain to accelerate commercialization. The development of a high-performance strain is still the greatest challenge that requires much time, effort, and money. Systems metabolic engineering and other advanced tools and strategies that have been continuously developed will play important roles in more rapidly and inexpensively developing industrial strains. Economical preparation of fermentable substrates from nonfood biomass remains to be advanced further. Instead of utilizing fermentable carbohydrates, the use of waste biomass and even carbon dioxide will be more actively pursued. Microbial fermentation mostly requires water, and thus the use and reuse strategy of fermentation water needs to be actively developed.

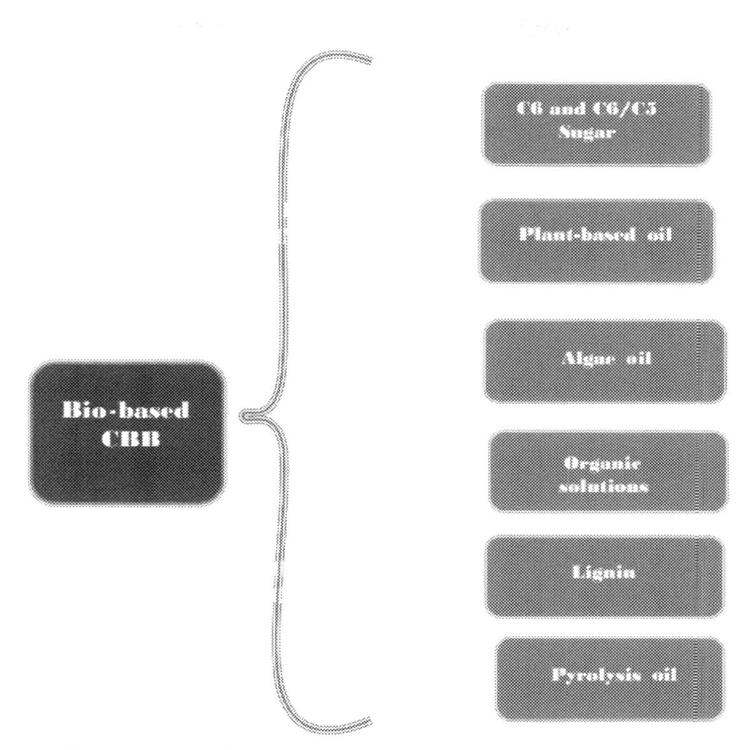

Figure 1.2 Classification of bio-based chemical building blocks.

Although such challenges still remain, it is great to see the enormous progress that has been made in the bio-based production of many of the platform chemicals. We are indeed moving toward a bio-based economy.

References

Corma, A., Iborra, S., Velty, A., 2007. Chemical routes for the transformation of biomass into chemicals. Chem. Rev. 107, 2411−2502.

Doran, P., 2012. Bioprocess Engineering Principles, 2nd edition Academic Press.

International Energy Agency, (2007), Bioenergy report task42: biorefinery, bio-based chemicals value added products from biorefineries.

Sierra, A.R., Zika, E., Lange, L., de Azua, P.L.R., Canalis, A., Esteban, P.M., et al., 2021. The bio-based industries joint undertaking: a high impact initiative that is transforming the bio-based industries in Europe. New Biotechnol. 60, 105−112.

United States Department of Energy Energy Efficiency and Renewable Energy, (2004) Report: top value added chemicals from biomass, volume i—results of screening for potential candidates from sugars and synthesis gas.

Wohlgemuth, R., Twardowski, T., Aguilar, A., 2021. Bioeconomy moving forward step by step−a global journey. New Biotechnol. 61, 22−28.

CHAPTER 2

Process intensification and sustainability

Contents

2.1 Process intensification and sustainability in bioblocks

The growing process of industrialization was a milestone for the world's economic evolution. Despite the contribution to the increase in quality of life, the global government policies remained far from the environmental impact that the growth of industrial activities could cause in our planet. The rapid increase in population resulted in increased food production with excessive industrialization, which led to increased pollution and resource depletion. In this way, natural resources began to be used as if there were no consequences on the environmental problem (Tobiszewski et al., 2009).

The concept of greening chemistry developed in the business and regulatory communities as a natural evolution of pollution prevention initiatives. In the efforts to improve crop protection, commercial products, and medicines, it also caused unintended harm to the planet and humans.

By the mid-20th century, some of the long-term negative effects of these advancements could not be ignored. Pollution choked many of the world's waterways and acid rain deteriorated forest health. There were measurable holes in the earth's ozone. Some chemicals in common use were suspected of causing or directly linked to human cancer and other adverse human and environmental health outcomes. Many governments began to regulate the generation and disposal of industrial wastes and emissions. The United States formed the Environmental Protection Agency (EPA) in 1970, which was charged with protecting human and environmental health through setting and enforcing environmental regulations.

Improvements in Bio-Based Building Blocks Production Through Process Intensification and Sustainability Concepts
DOI: https://doi.org/10.1016/B978-0-323-89870-6.00001-9

Green chemistry takes the EPA's mandate a step further and creates a new reality for chemistry and engineering by asking chemists and engineers to design chemicals, chemical processes, and commercial products in a way that, at the very least, avoids the creation of toxics and waste (Török and Dransfield, 2017).

We can develop chemical processes and earth-friendly products that will prevent pollution in the first place. Through the practice of green chemistry, we can create alternatives to hazardous substances. We can design chemical processes that reduce waste and reduce demand on diminishing resources. We can employ processes that use smaller amounts of energy. We can do all of this and still maintain economic growth and opportunities while providing affordable products and services to a growing world population.

Sustainable and green chemistry in very simple terms is just a different way of thinking about how chemistry and chemical engineering can be done. Over the years different principles have been proposed that can be used when thinking about the design, development, and implementation of chemical products and processes. These principles enable scientists and engineers to protect and benefit the economy, people, and the planet by finding creative and innovative ways to reduce waste, conserve energy, and discover replacements for hazardous substances.

It's important to note that the scope of green chemistry and engineering principles go beyond concerns over hazards from chemical toxicity and include energy conservation, waste reduction, and life-cycle considerations such as the use of more sustainable or renewable feedstocks and designing for end of life or the final disposition of the product (Török and Dransfield, 2017).

Green chemistry can also be defined through the use of metrics. While a unified set of metrics has not been established, many ways to quantify greener processes and products have been proposed. These metrics include ones for mass, energy, hazardous substance reduction or elimination, and life-cycle environmental impacts (Jiménez-González et al., 2012). In addition to these green metrics, there are a couple of fundamental aspects that must be considered together with this type of metric. A key factor is the use of energy for production. Another key aspect is the prevention of industrial waste materials, thus avoiding a greater environmental impact than that used by the operational activities of the industry (Lozano et al., 2018). In summary, the beginning of the 21st century has been markedly characterized by increased environmental awareness and pressure from legislators to curb emissions and improve energy efficiency by adopting greener technologies. In this context, the need for the chemical industry to develop processes that are more sustainable or eco-efficient has never been so vital. The successful delivery of green, sustainable chemical technologies at industrial scale will inevitably require the development of innovative processing and engineering technologies that can transform industrial processes fundamentally and radically. In bioprocessing, for example, genetic engineering of microorganisms will obviously play a major part in the efficient use of biomass, but the development of novel reactor and separation technologies giving high reactor productivity and ultimately high-purity products will be equally important for commercial success. Process intensification (PI) can provide such sought-after innovation of equipment design and processing to enhance process efficiency. In this sense, PI has objectives associated with improving productivity, capacity, security, and flexibility; and jointly decreases the use of energy, waste, and operating costs which naturally will allow a simplification of the processes (Gómez Castro and Segovia Hernández, 2019). The use of PI concepts has been able to demonstrate its usefulness, for example, through PI methodology it has been possible to reduce in some case studies 25% of investment costs, 35% in operating costs, and 40% in space requirements costs compared to the conventional process (Segovia Hernández and Bonilla Petriciolet, 2016). On the other hand, some other authors point out that through PI philosophy is possible to obtain novel and more sustainable processes (Babi et al., 2016). In other words, there is an important connection between the generation of sustainable designs and PI (Bravo García et al., 2021).

The term "process intensification" was probably first mentioned in the 1970s by Kleemann et al. (1978) and Ramshaw (1983). Ramshaw, among others, pioneered work in the field of PI. What does PI mean? Over the last two decades, different definitions of this term were published. Cross and Ramshaw (1986) defined PI as follows: "Process intensification is a term used to describe the strategy of reducing the size of the chemical plant needed to achieve a given production objective." In a review of PI, Stankiewicz and Moulijn (2002) proposed: "Any chemical engineering development that leads to a substantially smaller, cleaner, and more energy-efficient technology is process intensification." The BHR Group describes PI as follows (Ponce-Ortega et al., 2012): "Process Intensification is a revolutionary approach to process and plant design, development and implementation. Providing a chemical process with the precise environment it needs to flourish results in better products, and processes which are safer, cleaner, smaller, and cheaper. PI does not just replace old, inefficient plant with new, intensified equipment. It can challenge business models, opening up opportunities for new patentable products and process chemistry and change to just-in-time or distributed manufacture." Arizmendi-Sánchez and Sharratt (2008) highlighted two design principles for PI: (1) synergetic integration of process tasks and coupling of phenomena, and (2) targeted intensification of transport processes. Lutze et al. (2010) extended these principles by "adding/enhancing phenomena in a process through the integration of operations, functions, phenomena or through the targeted enhancement of phenomena in an operation" to stress the importance of phenomena-based thinking in PI.

To bring forward PI, Ponce-Ortega et al. (2012) expanded the meaning of the concept PI in a more holistic perspective: "PI as any activity aiming at the following five outcomes: (1) smaller equipment for given throughput; (2) higher throughput for given equipment size or given process; (3) less holdup for equipment or less inventory for process of certain material for the same throughput; (4) less usage of utility materials, and feedstock for a given throughput and given equipment size; and (5) higher performance for given unit size."

Furthermore, during the PI design, the relevance of the scale level (molecular scale, mesoscale, and macroscale) has been emphasized by describing four approaches, associated with the following domains: spatial, thermodynamic, functional, and temporal (Stankiewicz et al., 2019):

1. The spatial domain refers to the importance of well-defined structures or environments that allow maximization of process synergy (i.e., activity and selectivity of a certain catalyst are highly dependent on its geometrical structure).
2. The thermodynamic domain is focused on the optimum use of energy, considering alternative forms of energy and/or ways for its transfer.
3. In the functional domain, PI looks to maximize the synergy from different units, processes, or forms of energy, bringing multiple functions within a single component (i.e., reactive distillation units).
4. The temporal domain considers the manipulation of time scales of different processes in order to significantly reduce the processing times or process periodicity, for example, by using multitasking units or by switching from batch to continuous processing.

It is important to emphasize that during the synthesis of intensified process, the challenge is not only centered on the design of compacted units but on how to guarantee their functionality. Taking into account process objectives, chemical processing can be described by the elementary process function of each unit operation, or by the related process phenomena. Lutze et al. (2013) classified the different phenomena into eight categories: (1) mixing, (2) phase contact, (3) phase transition, (4) phase change, (5) phase separation, (6) reaction, (7) energy transfer, and (8) stream dividing. In summary, from its basic definition, PI is characterized by the development of novel equipment, wherein miniaturization and multifunctionality are fundamental targets. PI equipment can be divided into two categories: (1) equipment involving chemical reactions, and (2) equipment for operation not involving reactions.

As a complementary perspective, Portha et al. (2014) proposed classifying PI in two main categories: (1) local intensification, and (2) global intensification. Local intensification aims to use PI techniques to improve the efficiency of a single unit in isolation from the whole process. On the other hand, global intensification considers the performance of the whole process, taking into account the interactions among all units within the system.

As has been previously established, concerning green chemistry, Gómez Castro and Segovia Hernández (2019) have noted that the scope of PI is becoming much wider, considering the integration of not only heat and power but also water, safety, and other aspects of processes. Environmentally, the most telling impact of PI is likely to be in the development of reactor designs; since the reactor dictates both the product quality and the extent of the downstream separation and treatment equipment, during any chemical process. Thus, the development of reactor designs enables PI to be beneficial for energy use, impacting on the environment, through the delivery of a high quality product, without extensive downstream purification sequences.

All of the above, in recent years, PI has attracted considerable academic interest as a potential means for process improvement, to meet the increasing demands for sustainable production. PI is gaining much attention as one of the key objectives in designing new plants and retrofitting existing units. Several drivers have contributed to this increasing attention. For instance, enhanced process safety and homeland security are tied to PI, as the inventory and flows of hazardous substances are lowered, the process risk is typically reduced. As many processes, particularly those in the chemicals, nuclear and oil industries, involve the production, handling, and use of hazardous substances, PI is one way in which the inventory of such substances, and the consequences of a process failure, may be significantly reduced. PI, therefore, has the potential to be a significant factor in the implementation of inherent safety. Additionally, conservation of natural resources (including better utilization of mass and energy), biotechnological applications, new control methods may be linked to PI (Segovia Hernández and Bonilla Petriciolet, 2016).

In the case of biotechnological applications, bioprocess intensification represents one of the research focuses especially because approximately 50% of the elements used to decrease dependency on fossil feedstock can be obtained from renewable sources. Then, there is a tremendous need to develop bioprocesses that further optimize biomass harnessing through reaction and advanced separation techniques relying on sustainable ideologies. Additionally, bioprocesses face relevant challenges to be overcome in order to provide a sustainable future. Amongst the milestones identified within the international project Delft Skyline Debates for Process Intensification, the ones related to bioprocesses are (Górak and Stankiewicz, 2011):

- low-cost small-scale processing technologies for production applications in varying environments;
- recycling of composite materials: design, engineering, and intensified production technologies;
- toward perfect reactors: gaining full control of chemical transformations at the molecular level;
- elemental sustainability: toward the total recovery of scarce elements;
- production systems for personalized medicine;
- biohybrid organs and tissues for patient therapy;
- chemicals from biomass—integrated solution for chemistry and processing.

It is foreseen that achieving those milestones through bioprocess intensification, breakthrough technologies will address not only processing issues but crucial societal problems, such as human health, the availability of water and food, energy and material resources, transport, and living standards.

Bioprocesses intensification cannot be seen as a single area, because this has been developed from different perspectives considering multidisciplinary and multiscale aspects, allowing improvements in production, purification, and in overall performance. Bioprocesses have been intensified in different areas (e.g., genetic engineering, biology, biotechnology, chemical engineering, material science, etc.), the synergy among them, and the analysis at different levels of abstraction by a multiscale approach (from a cell manipulation up to lab/pilot plant scale). In order to accomplish a bioprocess intensification, the different areas have to interact and also have feedback for further improvements. For instance, the product yield in a biological system have been enhanced by increasing biocatalytic action (performed by genetic engineers and biologists), resulting in better substrate utilization (executed in the biotechnology and biochemical area) and have also been intensified by improving the reaction environment adapting other technologies (especially by chemical engineers). Starting from the core, there is a trend in bioprocess intensification for metabolizing multiple carbon sources to enhance substrate utilization (such as sugars, alcohols, etc.) by the same microorganism. This issue has originated the exploration of new manners to boost the performance of the biological entity by internal manipulation (i.e., microorganism mutation and genetic manipulation to adapt part of the metabolic route from a different microorganism). Enhancing the biocatalytic action has also allowed getting more robust microorganisms toward inhibitors and reaction conditions, letting to intensify product yield and productivity (Segovia Hernández and Bonilla Petriciolet, 2016).

The developments in the microorganism's level have allowed to attempts for improving the substrate utilization at the lab and reactor level. Different strategies for intensifying the reactor performance have been analyzed. This includes fermentation of one or more carbon sources with multiple microorganisms in the same bioreactor (so-called cocultivation), and the fermentation of two or more carbon sources by one microorganism (so-called cofermentation) instead of individual reactors for pentose and hexose in an ethanolic fermentation, for instance. The simultaneous saccharification and fermentation have been another intensification approach for substrate conversion where the enzymatic hydrolysis and fermentation of one of the liberated carbon sources from the water-insoluble solids is carried out simultaneously, instead of separate hydrolysis and fermentation. Another approach is the simultaneous saccharification and cofermentation with the synchronized enzymatic hydrolysis and fermentation of two or more liberated carbon sources in the same unit, rather than separate hydrolysis and cofermentation. There is another manner to intensify the process known as consolidated bioprocessing where the microorganism produces the necessary enzymes to liberated the carbon sources which are fermented by the same microorganism in the desired bioproduct (Segovia Hernández and Bonilla Petriciolet, 2016).

Even though the previous bioprocesses intensification has been achieved specifically from the reaction point of view, there is a top layer approach that includes the modification of the reaction environment. This layer refers to the in situ removal of the products obtained in the biochemical reaction. For certain bioprocesses, some of the products can impair the microorganism(s), thus, decreasing the productivity and product yield. In situ product removal/recovery investigates how to adapt external or internal devices to bioprocesses that permit to in situ remove inhibitors from the reacting vessel (i.e., main or secondary products). There is a special interest in how to integrate bioreactors with membrane separation processes such as micro/ultra/nanofiltration, pervaporation, membrane distillation, ion exchange membranes, and liquid membranes. Beyond those, other approaches have also been considered as the use of the removal/separation based on phase equilibria, such as liquid–liquid equilibria, stripping removal, and so on (Gómez Castro and Segovia Hernández, 2019).

In summary, there is an existing market for chemical building blocks, produced from bioprocesses, but it can be considered relatively immature, with development levels varying according to the building block considered and ranging from proof of concept in the laboratory to full commercial production. Strong cooperation within the value chain from feedstock producer to end-user is required for new chemical building blocks to successfully enter the market. In 2013, the demand for bio-based chemical building blocks in the EU was 1029 millions of euros (MEUR), equivalent to 35% of the total global production. The market grew at a compound annual growth rate of approximately 18.6% per annum. A combination of high feedstock, conversion and downstream processing costs means that the cost of producing bio-based chemicals is currently more expensive than processes using fossil fuel feedstocks. Opportunities for bio-based premiums to overcome price differentials for bio-based chemical building blocks should be considered. For example, the intensification of processes can be a strategy that based on minimizing the number of equipment with the maximum of unit operations, decreasing plant sizes and energy consumption will minimize the cost of production and the sale price of bio-based chemical building blocks.

The transition from carbon to biomass-based products is creating an unprecedented demand for bio-based chemical building blocks, and in turn a hard process engineering challenge, given the decreasing grade of the accessible oil, the need to address new and more complex industry, based in the use of green process, while reducing the environmental footprint. This challenge demands transformational change to the practice of production of chemical bioblocks, including the need for PI.

Finally, resources and the environment are two basic problems that all over the world encounter for economic development. As known, green chemical engineering is an important strategy to achieve sustainable development. In particular in the area of bio-based chemical building blocks. It is one of the important ways to solve the shortage of resources, energy, and alleviate environmental degradation. It is also the core foundation of science and technology for improving the quality of human life and ensuring green development, and will produce huge economic benefits. The intensification of the chemical engineering process, in the production of bio-based chemical building blocks, has the characteristics of "smaller, faster, safer, cheaper, more sustainable and cheaper." It is a new green chemical technology that can significantly reduce the size of the equipment. Furthermore, it has advantages of high efficiency, energy saving, and clean, therefore can promote green sustainable development in the area of chemical bioblocks. Interestingly, many PI initiatives designed to increase productivity also reduce the environmental footprint, which not only improves process and product profitability, but also overall sustainability in the industry of chemical products based in biomass (Stankiewicz et al., 2019). Accordingly, and by providing more cost savings, less energy consumption, emissions and pollutants reduction, higher-quality products with safety constraints consideration, the implementation of PI in the bio-based chemical building blocks industry can drastically provide the solution.

References

Arizmendi-Sánchez, J.A., Sharratt, P.N., 2008. Phenomena-based modularisation of chemical process models to approach intensive options. Chem. Eng. J. 135, 83.

Bravo García, J., Huerta Rosas, B., Sánchez Ramírez, E., Segovia Hernández, J.G., 2021. Sustainability evaluation of intensified alternatives applied to the recovery of nylon industry effluents. Process Saf. Environ. Prot. 147, 505–517.

Cross, W.T., Ramshaw, C., 1986. Process intensification—laminar flow—heat transfer. Chem. Eng. Res. Des. 64, 293.

Babi, D.K., Cruz, M.S., Gani, R., 2016. Fundamentals of Process Intensification: A Process Systems Engineering View, in Segovia Hernández, J.G., Bonilla Petriciolet, A., Process Intensification in Chemical Engineering: Design, Optimization and Control, First Edition Springer-Verlag.

Gómez Castro, F.I., Segovia Hernández, J.G., 2019. Process Intensification: Design Methodologies, first ed. De Gruyter.

Górak, A., Stankiewicz, A., 2011. Research Agenda for Process Intensification—Towards A Sustainable World of 2050. Institute for Sustainable Process Technology, Amersfoort, The Netherlands.

Jiménez-González, C., Constable, D.J.C., Ponder, C.S., 2012. Evaluating the "greenness" of chemical processes and products in the pharmaceutical industry – a green metrics primer. Chem. Soc. Rev. 41, 1485.

Kleemann, G., Hartmann, K., Wiss, Z., 1978. Tech. Hochschule "Carl Schorlemmer,". Leuna Merseburg 20, 417.

Lozano, F.J., Lozano, R., Freire, P., Jiménez-Gonzalez, C., Sakao, T., Ortiz, M.G., et al., 2018. New perspectives for green and sustainable chemistry and engineering: approaches from sustainable resource and energy use, management, and transformation. J. Clean. Prod. 172, 227.

Lutze, P., Gani, R., Woodley, J.M., 2010. Process intensification: a perspective on process synthesis. Chem. Eng. Process. 49, 547.

Lutze, P., Babi, D.K., Woodley, J.M., Gani, R., 2013. Phenomena based methodology for process synthesis incorporating process intensification. Ind. Eng. Chem. Res. 52, 7127.

Ponce-Ortega, J.M., Al-Thubaiti, M.M., El-Halwagi, M.M., 2012. Process intensification: new understanding and systematic approach. Chem. Eng. Process. 53, 63.

Portha, J.F., Falk, L., Commenge, J.M., 2014. Local and global process intensification. Chem. Eng. Process. 84, 1.

Ramshaw, C., 1983. HiGee' distillation—an example of process intensification. Chem. Eng. 389, 13.

Segovia Hernández, J.G., Bonilla Petriciolet, A., 2016. Process Intensification in Chemical Engineering: Design, Optimization and Control, First Edition Springer-Verlag.

Stankiewicz, A., Moulijn, J.A., 2002. Process intensification. Ind. Eng. Chem. Res. 41, 1920.

Stankiewicz, A., Van Gerven, T., Stefanidis, G., 2019. The Fundamentals of Process Intensification, first ed. Wiley-VCH.

Tobiszewski, M., Mechlińska, A., Zygmunt, B., Namieśnik, J., 2009. Green analytical chemistry in sample preparation for determination of trace organic pollutants. TrAC Trends Anal. Chem. 28, 943.

Török, B., Dransfield, T., 2017. Green Chemistry—An Inclusive Approach, first ed. Elsevier.

CHAPTER 3

Basic concepts on simulation of (bio)chemical processes

Contents

3.1 (Bio)chemical processes

For society today, biomass is undoubtedly a valuable renewable resource for the production of biomaterials and bioenergy. Due to great growth in the biotechnology, process engineering, and genetics sectors, as well as others, novel ideas have emerged for the generation of new products from the conversion of biomass. By their nature, the processes for the generation of biomaterials and bioenergy are complex. These processes are usually represented in terms of mathematical equations. However, the degree of complexity of the process, in turn, can make mathematical representation complicated, to the point that its resolution is impossible analytically. Added to this, the current trend does not simply address the manufacture of a certain biochemical, but this process must be in its optimal operating conditions, meet market demands, and handle market uncertainties. It is clear that in the biomaterials and bioenergy sector, because of the complexity of its processes, the use of process simulation is a priority.

3.2 Concept of simulation in bioprocesses (chemical)

The concept of process simulation extended to the area of biomaterials and bioenergy is nothing more than the representation of the process in

Improvements in Bio-Based Building Blocks Production Through Process Intensification and Sustainability Concepts
DOI: https://doi.org/10.1016/B978-0-323-89870-6.00007-X

terms of a set of mathematical equations, which is then resolved to obtain information on the performance of the process. Korn (2007) said, "Simulation is experimentation with the models." Because of the complexity of chemical and biochemical processes, the mathematical model is usually solved with computational tools using simulators. Finding a solution involves several parameters and assimilates the effects of variables. Previously, it was enough to find solutions where the requirements of purity and economics of the desired product were met. Today, with powerful simulation tools, conditions where the set of variables satisfy more than one objective function can be found. Thus, optimizing the process in terms of the desired indices (economic, environmental impact, inherent safety, etc.) to obtain sustainable processes and as a circular economy is required.

The design and the optimization of the biochemical processes are highly nonlinear problems involving continuous and discrete design variables. Within the simulation and optimization of each bioprocess, a solution is given to material and energy balance, equilibrium equations, complex reaction kinetics, the set of equations "MESH" (Material balance, Equilibrium, Summation and Heat), and so on. For this reason, the use of simulators and computational tools are of great value in this area. Furthermore, the objective functions are potentially nonconvex, with the possibility of finding local optimum, subject to constraints.

Simulators generally have two types of basic structures: fixed-structure and variable-structure. Fixed-structure simulators are those in which the correspondence of flows with the unit modules and the path of calculations through the program is determined by the input data provided by the user, who obtains results based on these calculations. In variable-structure simulators, a different executive model is written or generated for each unit and process to be studied.

3.2.1 Simulation categories for biochemical processes

There are several simulation classifications, and these depend on the model used (see Fig. 3.1). It should be noted that almost all of simulation tools to be discussed in this chapter are applicable throughout the life cycle of the bioproduct and bioprocess. It is therefore difficult to limit a rigorous discussion on simulation, design, and optimization. The tools described in this chapter are described in later sections because, for example, dynamic simulators are valuable tools only for the specific types of tasks to be discussed here. That is why this chapter will address simulation

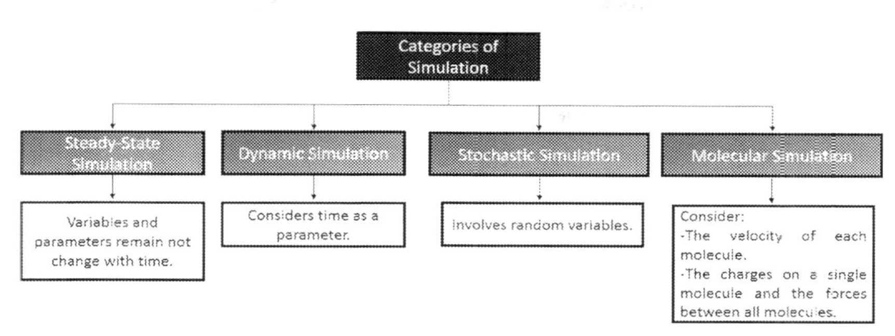

Figure 3.1 Categories of simulation.

in the stationary and dynamic states, all referring to the area of biomaterials and bioenergy. The types of equations, variables, problem formulation, and resolution, in various types of simulation, turn out to be different depending on the situation.

3.2.1.1 Steady-state simulation

In the area of bioprocesses, it is very common to encounter simulations in continuous mode. That is, both parameters and variables do not change depending on time. In other words, on the balance sheets, energetic, and, for the time being, the term accumulation is zero. Steady-state models are usually simple, with no changes regarding time. In a steady-state simulation, process topology, design and operating parameters of the units, and composition of the products are generally found. However, steady-state simulations are not able to predict the dynamic behavior of the process (Suckow, 2016).

There is a variety of simulators to reproduce biochemical processes in a steady state. The most prominent fixed-structure simulators include Aspen Plus and Aspen Hysys (Aspen Technology, Inc.), Chemcad (Chemstations, Inc.), and PRO/II (Simulation Sciences). Aspen Plus, Chemcad, and PRO/II are sequential modular simulators, while Aspen Hysys uses a simultaneous modular algorithm. In variable-structure simulators, several programming languages are useful in the area of bioprocesses, including Matlab (MATrix LABoratory), gPROMS (general PROcess Modeling System), GAMS (General Algebraic Modeling Language), FORTRAN (Formula Translating System), Julia Language, Python, and so on.

3.2.1.2 Dynamic simulation

Within the industry of biochemical processes, several types of equipment and processes that require a continuous, discontinuous, or semicontinuous operation can be found. Dynamic simulation has an equal interest in continuous processes, such as in already inherently dynamic operations such as discontinuous or semicontinuous ones. Dynamic processes assist both in the design of process control systems, in the startup and shutdown of a process, since this cannot be studied by steady-state simulations. When you view a simulation over time as another parameter of the process, that is known as a dynamic simulation. Dynamic simulation applications range from achieving process design under optimal operating conditions and time to designing and operating processes operationally feasible, safe, economic, and sustainable. That is why dynamic simulation has become an indispensable tool to use in design processes and controllers alike. Control strategy and controller tuning is usually the main objective of dynamic simulation, but without proper control system configuration, the model will end up in completely abnormal operating conditions. That is to say, dynamic simulation becomes transcendental in bioprocesses because, in the instrumentation of the process and when choosing the type of controller (Proportional-Integral-Derivative (PID)being the most common), it is oriented to the optimal performance of the process.

Unlike the steady-state model, the implementation time of a dynamic model is usually longer than the time required to implement a steady-state model (Mokhatab and Poe, 2012). Therefore, commercial simulators representing dynamic status are used, which allows the development of dynamic models quite easily. Rigorous dynamic simulation can be performed using simulators such as Aspen Plus Dynamics, Aspen Hysys Dynamics, or DynSim.

3.2.2 Process simulation biochemical applications

The applicability of simulation tools in the area of biochemical processes is very wide. It covers the design, operation and commissioning, control, and optimization of the bioprocess. As mentioned earlier, simulations are based on the reproduction of any process through mathematical modeling. By its complex nature (material and energy balance, equilibrium equations, reaction kinetics, presence of azeotropes, and the set of equations "MESH"), modeling these bioprocesses turns out to be a complicated task. It starts from the reproduction of pretreated, fermentation, separation and purification schemes, and so on, which are highly nonlinear and difficult to solve. Therefore, the use of simulation tools in the bioprocesses sector turns out to be fundamental in the study of new bioprocesses, as well as in the

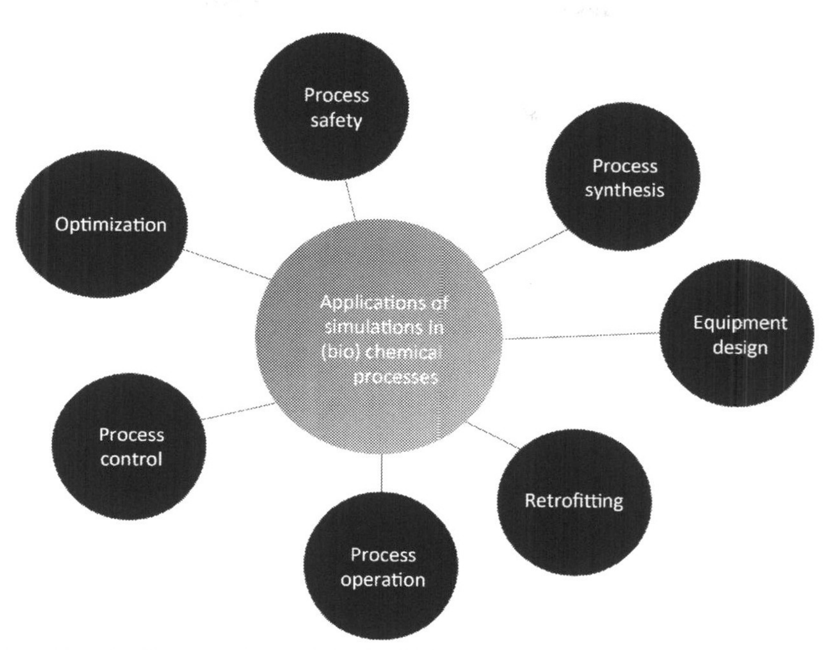

Figure 3.2 Applications of simulation in (bio)chemical processes.

improvement of existing ones. It would be almost impossible to carry out the design and/or optimization of a bioprocess experimentally, due to the time and effort to change the process variables and observe the effects of the process. Fig. 3.2 shows some of the broad areas of application where simulation of bioprocesses plays an important role (Mokhatab and Poe, 2012). This chapter will address some of them in specific ways.

3.2.2.1 Synthesis and process design biochemicals

The synthesis and design of any process are generally well-established procedures in the biochemical industry. Process synthesis is the stage in the design where components are selected and how to interconnect them to create your flowchart. All industrial processes are born from previous studies of economic profitability, environmental impact, inherent safety, and so on. The way to transfer information and generate a new bioprocess ranges from the provision of experimental data, the simulation of the bioprocess, the use of pilot plants, to the construction of an industrial plant. The simulation stage plays a very large role in the design of an industrial plant. That is to say, if at the testing stage the bioprocess turns out to be economically viable, one can move on to the next phase. On the other hand, if the bioprocess is not feasible, then the process design is discarded. In addition, a bioprocess may be cost-effective but operate with low efficiency, and that is where simulation helps to find the right variables and thus increase the efficiency of the bioprocess. In other words, in the synthesis and design stage of a bioprocess, simulation is elementary to perform a first sketch of the process, and then with the help of optimization tools have the optimal design of the bioprocess. Simulators are also of great help in terms of seeking the intensification of a bioprocess.

Various types of software can be used currently for the design and intensification stage of a bioprocess. Among the most popular are Aspen Plus, Aspen Hysys, Chemcad, PRO/II, UNISIM, and so on. Saving time and effort in process design is of paramount importance in light of tough competition and rapid technological progress.

3.2.2.2 Operation, control, and safety of processes biochemicals

While having the design of a biochemical process is a great development, keeping the process running and maximizing profits is even more so. Biochemical processes are characterized by fluctuations in the raw materials used, as some turn out to be stationary and may cease to exist for periods of the year. Uncertainty in the availability and price of raw materials, the price of the bioproduct, and demand for the bioproduct all affect the optimal operating conditions of the process. The simulation helps to quickly estimate the optimal operating conditions of the process and determine the feasibility of the process in front of a variety of scenarios.

Of equal importance, the quality and performance of the bio product generated should be taken into consideration. For example, a biofuel has quality standards in order to be able to be brought to market. Controlling a process requires the implementation of the process. That is controllers, measuring instruments, valves, and so on. Because of the complex nature of bioprocesses and disturbances, dynamic behavior cannot always be expressed analytically. This is where simulation helps to obtain the best control properties through rigorous models to be able to choose from a range of designs generated in terms of control. Within the range of software for operation simulation and process control, you can find Aspen Plus Dynamics, Aspen Hysys Dynamics, or DynSim.

At the same time, simulation should be used to analyze various aspects concerning hazards and ways of handling them. In other words, in bioprocesses, there are various risks associated with explosions, toxicity, component leakage, and so on, and that is where simulation supports in the design and evaluation of bioprocesses and the reduction of environmental pollution.

By generating a bioprocess with all of these simulation tools in economic, environmental, inherent safety, and control terms, a sustainable process is automatically being generated and can be delineated dependent on the case in the concept of the circular economy.

3.3 Concept of modeling and tools in process biochemicals

Various simulators already mentioned in this chapter (Aspen Plus, Aspen Hysys, Chemcad, etc.) have models of various equipment, thermodynamic properties of multiple components, and numerical methods for their resolution. The resulting design in such simulators depends on the input data provided to it. In the case of other simulators, it is necessary to generate a robust model, since the success of the simulation will depend on the model. Today, programming languages for generating robust models have been simplified. There is a wide variety of software to generate models in the area of bioprocesses, including Matlab, gPROMS, GAMS, FORTRAN, Julia Language, Python, and so on. Knowing the syntax and semantics of each of the aforementioned software, you can develop the bioprocess models. Recent graphical interfaces make model construction fast and less complicated. However, an optimal understanding of the process is required for model validation.

The model is supposed to be able to represent and predict the behavior of a bioprocess. When solving the model, and with the results obtained, the process must be analyzed under various edges. It has been pointed out on several occasions that the representation of a system by a model does not capture reality. It is limited to a certain perspective of reality that is considered relevant in the context in which the model is supposed to be used (Klatt and Marquardt, 2009).

3.4 The role of simulation and process modeling biochemicals

Software tools that model and simulate biochemical processes from a steady-state perspective, or dynamic simulation, are becoming increasingly important. Currently, these tools offer sophisticated graphical interfaces, which greatly improve their use. The main role of simulation and modeling in biochemical processes is that they represent the transport and transformation of chemical species, as well as separation and purification processes until the final product is obtained. By performing good modeling and/or simulation of a bioprocess, designs can be generated that in the end mean real industrial designs, and that bring society closer to a transformation to bioproducts, with a character of sustainability and circular economy.

3.5 The role of process optimization biochemicals

The search for bioprocesses in their operating conditions and optimal design parameters to meet economic, environmental impact, safety, control, and so on is increasingly required to give a sustainable character to the processes. In other words, the bioprocess industry (biotechnology, food, pharmaceutical, environmental, etc.) aims to increase its productivity, profitability, and/or efficiency, without neglecting aspects of green chemistry. Therefore, optimization has become a fundamental tool for optimally designing and operating production facilities in these sectors.

In recent years, robust and efficient optimization techniques have been developed that can be used, in combination with suitable models, to obtain optimal or almost optimal solutions in the design, operation, and control of bioprocesses. Due to the high nonlinearity of the models, the rigorousness of the simulation, and the search for multiobject optimization, the most appropriate optimization tools will be used for each case. Therefore, the following chapters will present each of the optimization tools used for each of the bioproducts, which will ensure that the proposed intensified processes meet the requirements of green chemistry, sustainability, and the circular economy.

References

Klatt, K.U., Marquardt, W., 2009. Perspectives for process systems engineering—personal views from academia and industry. Comp. Chem. Eng. 33 (3), 536–550.

Korn, G.A., 2007. Advanced Dynamic-system Simulation: Model-replication Techniques and Monte Carlo Simulation. John Wiley & Sons.

Mokhatab, S., Poe, W.A., 2012. Handbook of Natural Gas Transmission and Processing. Gulf Professional Publishing, pp. 467–470.

Suckow, M., 2016. Process modelling and simulation in chemical, biochemical and environmental engineering. Von AK Verma. Chem. Ing. Tech. 88 (12), 1975.

CHAPTER 4

Bioethanol

Contents

4.1 Bioethanol

Bioethanol is listed as the ideal substitute for gasoline on land transport vehicles. Bioethanol, also known as ethyl alcohol (C_2H_5OH) is a transparent and colorless liquid of low toxicity, biodegradable that causes little environmental pollution (Rutz and Janssen, 2007). Bioethanol is a fuel that offers advantages by virtue of its physicochemical characteristics. Some of those characteristics being a low-density liquid with high fluidity and with a high combustion heat. Above all, bioethanol is superior in environmental terms due to its sustainability and low environmental effects.

Bioethanol was also discovered to be a high-octane fuel and has therefore replaced lead as an octane booster in gasoline. Bioethanol mixtures are commonly presented with gasoline in order to oxygenate the fuel mixture, thus burning it completely and reducing polluting emissions. These bioethanol fuel blends are commonly offered with 10% bioethanol and 90% gasoline (E10). An advantage to this is that neither vehicle

Improvements in Bio-Based Building Blocks Production Through Process Intensification and Sustainability Concepts
DOI: https://doi.org/10.1016/B978-0-323-89870-6.00005-6

engines require an alteration to run with E10 nor do vehicle warranties seem to be affected by it. However, not only can the E10 be found on the market, but other flexible fuel vehicles that run on up to 85% ethanol and 15% gasoline blends (E85) are available (Sindhu et al., 2019).

The potential use of bioethanol is not limited to the field of biofuels; it can also be used as a versatile raw material in various industries, such as the pharmaceutical industry, the cosmetic industry (astringent), the chemical industry (as a compound of starting in the synthesis of various products) and it is an excellent solvent, antifreeze, and disinfectant (Evans, 1997).

It is clear, from its various areas of application that the production and consumption of biofuels such as bioethanol are increasing. In 2015 the world production of bioethanol fuel was valued at more than 97 billion liters, with 85% of it being produced in two countries: the United States with 56 billion liters and Brazil with 27 billion liters (Alternative Fuels Data Center, 2020). In Japan, the national goal for 2025 is the introduction of bioethanol to supply the equivalent of 500 billion liters of oil. However, to meet the national biofuel target, Japan would have to depend largely on imports due to the high cost of producing bioethanol in Japan. Simultaneously, countries such as Australia, Belgium, Canada, Italy, Norway, and the United States have already implemented biofuel blending mandates (Yamamoto, 2017).

With all the momentum from global policies, the size of the global bioethanol market has been increasing. In 2019 the bioethanol market was valued at $89 trillion, with a compound annual growth rate of 4.8% expected in the period 2020–27 (GVR, 2020). The demand for bioethanol is driven by the growing use of the product as a biofuel.

Although several bioethanol production plants have been built around the world, a great effort is still needed to reduce production costs. The principal limitations in the production of bioethanol are the high costs of raw material, of enzymes, of detoxification and the recovery of bioethanol.

Ethanol can be made from both natural and petrochemical raw materials. The following sections will show the conventional routes of ethanol production while highlighting its advantages and disadvantages.

4.2 Petrochemical route of ethanol production

4.2.1 Process, raw material, and kinetics

There are several routes for the production of ethanol, such as fermentation (bioethanol), the indirect hydration process (esterification–hydrolysis)

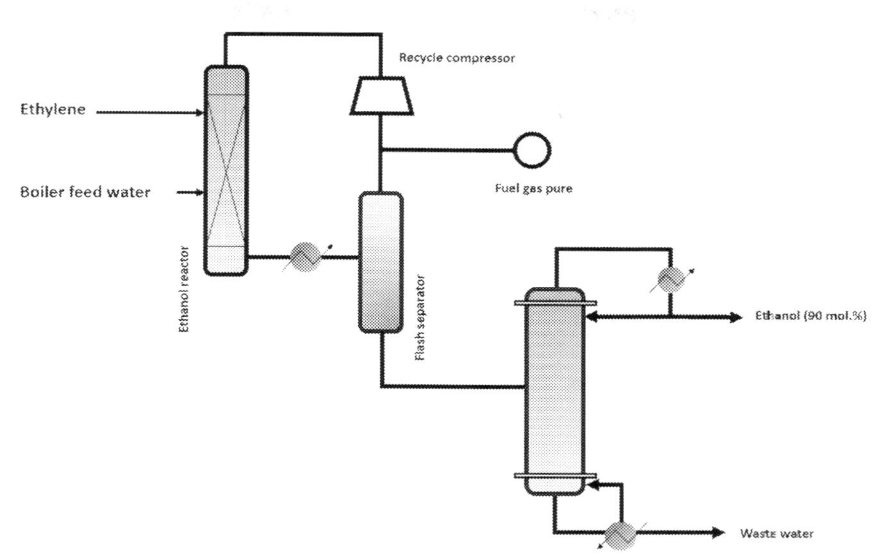

Figure 4.1 Ethanol production process through direct ethylene hydration.

and the direct hydration of ethylene. Since 1947, synthetic ethanol has been produced by the direct hydration of ethylene replacing the indirect hydration of ethylene. Shell (Weissermel and Arpe, 2008) first introduced the direct catalytic hydration of ethylene. Through this process, ethanol is produced by a chemical reaction of ethylene with water vapors. The reaction is reversible and exothermic. The complete scheme for the production of anhydrous ethanol (>99 mol.%) through direct hydration of ethylene, consists of three stages: (1) reaction, (2) recovery, and (3) purification (as can be seen in Fig. 4.1). The implicit reactions are shown in Eqs. 4.i–4.iii (Hidzir et al., 2014).

$$CH_2 = CH_2(g) + H_2O(g) \overset{r_1,r_2}{\leftrightarrow} CH_3CH_2OH(g) \tag{4.i}$$

$$C_2H_2 + H_2O(g) \overset{r_3}{\rightarrow} CH_3CHO \tag{4.ii}$$

$$2C_2H_5OH + H_2O(g) \overset{r_4,r_5}{\leftrightarrow} H_2O + (CH_3CH_2)_2O \tag{4.iii}$$

Eqs. 4.1–4.3 show each of the rates of reaction (Loïc et al., 2019).

$$r = r_1 - r_2 = \frac{k_1 \cdot p_E \cdot p_W - k_2 \cdot K_A \cdot p_A}{(1 + K_E \cdot p_E + K_W \cdot p_W + K_A \cdot p_A + K_{DEE} \cdot p_{DEE})^2} \tag{4.1}$$

$$r = r_3 = k_3 \cdot p_{\text{Acetylene}} \tag{4.2}$$

$$r = r_4 - r_5 = \frac{k_4 \cdot p_{\text{A}}^2 - k_5 \cdot p_{\text{w}} \cdot p_{\text{DEE}}}{(1 + K_{\text{W}} \cdot p_{\text{W}} + K_{\text{A}} \cdot p_{\text{A}} + K_{\text{E}} \cdot p_{\text{E}} + K_{\text{DEE}} \cdot p_{\text{DEE}})^2}, \tag{4.3}$$

where r_1, r_2, r_3, r_4, r_5—rate of reaction, k_1, k_2, k_3, k_4, k_5—reaction constant, p—partial pressure, p_{w}—water partial pressure, p_{A}—acetaldehyde partial pressure, p_{E}—ethanol partial pressure, p_{DEE}—diethyl ether partial pressure, K—equilibrium constant, K_{W}—water equilibrium constant, K_{E}—ethanol equilibrium constant, K_{A}—acetaldehyde equilibrium constant, K_{DEE}—Diethyl ether equilibrium constant.

The kinetics show that with the Langmuir Hinshelwood type reaction mechanisms, it is possible to obtain the reaction rates of the gas phase hydration of ethylene, dehydration of ethanol, and conversion of acetylene to acetaldehyde. In this mechanism and with standardized operating conditions (pressure 70–80 atm, and temperature of 250°C–300°C), the limiting stage in the hydration of ethylene is the chemical reaction and not the diffusion mechanism. Thus, it is shown that the catalyst operates in a chemical regime and that it is considered isothermal (Loïc et al., 2019).

The direct hydration of ethylene has been carried out for more than 70 years in the chemical industry on catalysts consisting of silica gel with a high load of phosphoric acid. The catalyst is specified as being the supported liquid phase and the active catalyst attached to a vehicle such as the concentrated liquid acid (Fougret and Hölderich, 2001). Usually, a fixed bed reactor is used to carry out the direct hydration of ethylene, this is because the reagent is in a mobile fluid phase and the reaction takes place on the surface of the catalyst. Therefore, the reagent diffuses, adsorbs and reacts on the active surface of the catalyst (Qi et al., 1999). Loïc et al., 2019 show a maximum ethylene conversion of about 10%.

4.2.2 Performance index in the production of ethanol through petrochemical

Loïc et al., 2019 carried out an economic-energy analysis of a petrochemical ethanol production plant, executing heat integration (technical translation: heat integration) in the process shown in Fig. 4.1. They calculated the annual capital cost and the annual operating cost. The costs of raw materials were established using the market price. When establishing the total annual cost (approximately $12'780,000) it was observed that it was higher than both the income of the process and the construction of a new plant, thus making it not profitable at all. Moreover, when producing ethanol by hydrating ethylene, the energy consumption is 62 MJ/kg of ethanol produced; while the production of ethanol from natural raw materials only requires 19 MJ/kg of ethanol produced.

4.2.3 Disadvantages in the production of ethanol through petrochemical

However, the production of anhydrous ethanol through direct hydration of ethylene presents a series of disadvantages compared to its production through fermentation. The principal problems of this production route are associated with the stable production of anhydrous ethanol. Overall, good anhydrous ethanol yields are not presented, which makes its application on a commercial scale difficult (Jeong et al., 2012). Moreover, and due to its petrochemical nature, there are innumerable disadvantages compared to biological processes. In other words, when producing ethanol through direct hydration of ethylene, unfavorable effects can occur in our environment, such as air, water, and soil pollution. Petrochemical processes tend to cause loss of biodiversity and destruction of ecosystems. It is because of this that opting for a more environmentally friendly ethanol production route is an initiative with a greater future projection and to generate sustainable processes in this area.

4.3 Conventional bioethanol production process

Despite showing various advantages over ethanol production through petrochemicals, bioethanol production remains a complicated process. The transformation of biological sources such as cereal shells rich in sugar or lignocellulose requires the appropriate conditions and pretreatment of raw material to achieve fermentation and convert the organic matter into bioethanol. The fermentation product is a bioethanol solution with water that must be concentrated, and it subsequently has to be subjected to a suitable dehydration process. This, to be used as an oxygenator for gasoline (see Fig. 4.2).

4.3.1 Raw material for the production of bioethanol

Different raw materials, those that can easily be transformed into fermentable sugar, have been used for the production of bioethanol.

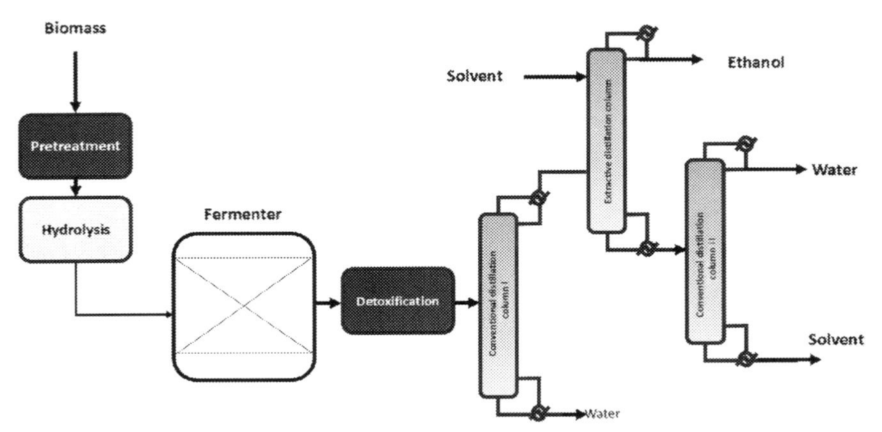

Figure 4.2 General scheme for obtaining bioethanol.

Its practical use will be determined by the yield in bioethanol by its cost and by the type of microorganism (Hernández-Nodarse, 2007). Various authors (Demirbaş, 2005; Balat, 2011; Talebnia et al., 2010), agree on a definition of three types of raw materials for the production of bioethanol:

1. simple sugar bearer materials that contain carbohydrates as a source of sugars: such as sugarcane juice, molasses, sweet sorghum, and so on;
2. starch materials, which contain starch as a source of sugars: such as cassava, corn, potato, and so on;
3. cellulosic materials, containing cellulose, hemicellulose: such as bagasse, wood, agricultural residues, and so on.

Fig. 4.3 describes various routes of bioethanol production through biomass. In this figure, the fundamental stages of each route depending on its precursor raw material can be seen.

The production of bioethanol from lignocellulosic biomass is an attractive alternative since lignocellulosic raw materials do not compete with food crops and are less expensive than conventional agricultural raw materials. Out of all the biomasses shown, lignocellulose is the most abundant. The biological conversion of various materials with a high lignocellulosic content (forest and agricultural residues, or lignocellulosic crops) offers numerous environmental benefits. However, this process is hampered by economic and technical issues (Sanchez and Cardona, 2008).

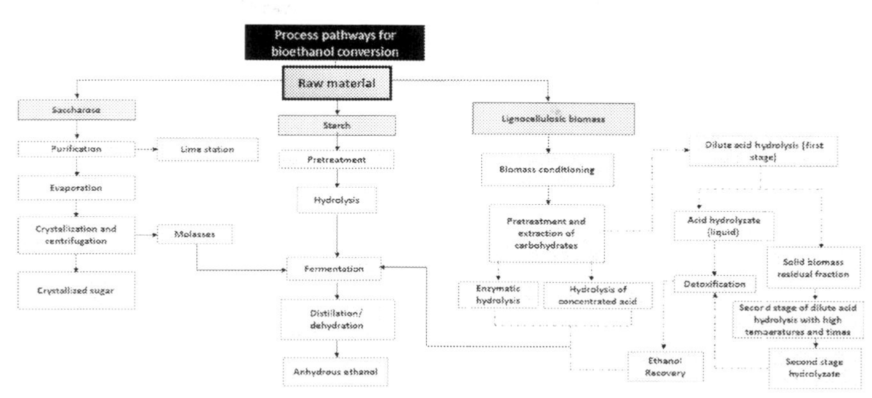

Figure 4.3 Ways of obtaining bioethanol.

4.3.2 Production of bioethanol from lignocellulosic biomass

The production of bioethanol from lignocellulosic biomass involves a series of stages such as pretreatment, enzymatic hydrolysis, detoxification, sugar fermentation, recovery, and purification of bioethanol to perform with the fuel specifications.

4.3.2.1 Pretreatment

The stage of hydrolyzing lignocellulose into fermentable monosaccharides continues to be a technical problem since the digestibility of cellulose is limited by many physicochemical, structural, and composition factors (Mosier et al., 2005). Due to these structural characteristics, a previous pretreatment stage has to be executed in order to obtain the sugars that will potentially ferment in the subsequent hydrolysis stage. The selection of the appropriate pretreatment technology for a particular feedstock depends on several properties. These properties are: high yields for multiple crops, site ages, harvest times, highly digestible pretreated solid, an insignificant degradation of sugars, operation in reactors of reasonable size and moderate cost, no production of solid waste, efficiency with low moisture content, the obtainment of high sugar concentration, fermentative compatibility, recovery of lignin, and so on (Alvira et al., 2010). Table 4.1 shows a series of pretreatments for the production of bioethanol from lignocellulose.

4.3.2.2 Enzymatic hydrolysis

Enzymatic hydrolysis is the process catalyzed by enzymes called cellulases, whose purpose is the degradation of cellulose. In the use of pretreatments,

Table 4.1 Biological, physical, chemical and physicochemical pretreatments to obtain bioethanol from lignocellulosic biomass (Alvira et al., 2010).

Pretreatments	
Biological pretreatments	
• Fungal pretreatment	Use of: *Phanerochaete chrysosporium*, *Ceriporia lacerata*, *Cyathus stercolerus*, *Ceriporiopsis subvermispora*, *Pycnoporus cinnarbarinus* and *Pleurotus ostreaus*.
Physical pretreatments	
• Mechanical comminution	Decrease of particle size and cristallinity of lignocellulosic in order to growth the specific surface and reduce the degree of polymerization.
• Extrusion	The constituents are subjected to heating, mixing, and shearing, resulting in physical and chemical variations during the passage through the extruder.
Chemical pretreatments	
• Alkali pretreatments	Growth cellulose digestibility and they are more effective for lignin solubilization, exhibiting insignificant cellulose and hemicellulose solubilization than acid or hydrothermal processes.
• Acid pretreatment	Solubilize the hemicellulosic fraction of the biomass and to make the cellulose more accessible to enzymes.
• Ozonolysis	Ozone is a powerful oxidant that shows extraordinary delignification efficiency.
• Organosolv	The highest advantage of organosolv process is the recovery of relatively pure lignin as a by-product.
Physicochemical pretreatments	
• Steam explosion: SO_2-steam explosion	The biomass is exposed to pressurized steam for a time ranging from seconds to several minutes, and then suddenly depressurized.
• Liquid hot water	Pressure is used to maintain water in the liquid state at elevated temperatures and incite alterations in the structure of the lignocellulose.

(*Continued*)

Table 4.1 (Continued)

Pretreatments	
• Ammonia fiber explosion	Biomass is process with liquid anhydrous ammonia at temperatures between 65°C and 100°C and high pressure for a flexible time. The pressure is then released, resulting in a rapid expansion of the ammonia gas that causes swelling and physical disturbance of biomass fibers and partial decrystallization of cellulose.
• Wet oxidation	Employs oxygen or air as catalyst. It permits reactor operation at comparatively low temperatures and short reactor times.
• Ultrasound pretreatment	Employed for extracting hemicelluloses, cellulose, and lignin.

the hydrolysis process is facilitated. At present, in addition to the benefits of sugars from cellulose, there is great interest in other sugars such as pentoses (derived from hemicellulose), which lead to the use of enzymes that act on said substances, for example with xylanases and xylases (Sánchez Riaño et al., 2010). According to Gómez Tovar (2008), the enzymatic hydrolysis for the cellulase complex can be summarized in three large groups:

1 endo β-glucanases: break the β-glucosidic bonds in a random way inside cellulose molecules;
2 exo β-glucanases: gradually attack cellulose molecules from nonreducing terminals releasing subunits of cellobiose;
3 β-glucosidases: hydrolyze low molecular weight cellobioses and cellodextrins (celotriose and celotetrose) into glucose.

Some microorganisms exist from which enzymes are obtained in order to carry out hydrolysis. Table 4.2 shows some of the microorganisms that can be obtained commercially (Ovando-Chacón and Waliszewski, 2005).

4.3.2.3 Detoxification

The purpose of detoxification is to eliminate substances that could be tentatively formed during the submission of raw material to pretreatment and enzymatic hydrolysis, which are toxic and inhibitory in fermentation. These substances are typically created due to the high temperatures and acidic conditions in which the previous stages develop. In the same way, it seeks to avoid the formation of other substances during the fermentation process

Table 4.2 Microorganisms from which hemicellulosic enzymes are obtained.

Type	Characteristics	Microorganisms
Fungus	Aerobics	*Trichoderma viride, Trichoderma reesei, Penicillium pinophilum*, and *Trichoderma koningii*
	Thermophilic aerobics	*Sporotrichum thermophile, Thermoascus aurantiacus* and *Humicola insolens*
	Mesophilic anaerobes	*Neocallimastix frontalis, Piromonas communis, Sphaeromonas communis*
Bacterium	Mesophilic and thermophilic aerobics	*Cellulomonas* Sp., *Cellvibrio* Sp., *Microbispora bispora*
	Mesophilic and thermophilic anaerobic	*Acetivibrio cellulolyticus, Bacteroides Succinogenes* and *Clostridium Thermocellum*

that affects ethanol production (Sánchez Riaño et al., 2010). According to Oliva Domínguez (2004), these substances are usually grouped into three categories: furan derivatives, low molecular weight aliphatic acids and phenolic derivatives. There is a range of possibilities to carry out detoxification, and these are classified according to the type of substances and actions that constitute them. Several authors group them into physical, chemical, and biological methods (Oliva Domínguez, 2004).

- Biological method: this method involves the use of specific enzymes and microorganisms that act on toxic compounds by changing their composition. In detoxification with enzymes are mainly the use of peroxidases and laccases. As for the microorganisms used for detoxification, they are both bacterial and fungal (Mussatto and Roberto, 2004).
- Physical and chemical methods: these methods consist of a series of techniques depending on the type of substance and whose action ranges from evaporation, extraction, treatment with hydroxides, the use of activated or vegetable carbon, the use of zeolites, advanced oxidation, and so on.

4.3.2.4 Fermentation

Fermentation is a process that can simply be defined as a chemical change caused by the action of microorganisms. Where the activity of some microorganisms that process sugars (glucose, fructose, sucrose, starch, etc.)

guide it to produce alcohol in the form of bioethanol, carbon dioxide (g) and ATP molecules that the microorganisms themselves consume in their anaerobic cellular energy metabolism. There are two key components in the fermentation process: the microorganism and the substrate. The technique of fermentation of lignocellulosic biomass to bioethanol demonstrates to be quite similar to any other fermentation process. The most commonly used fermentation process in the bioethanol industry is submerged fermentation. However, about the production of lignocellulosic bioethanol, two fundamental routes can be distinguished. The first route is through the separate processes of hydrolysis and fermentation (SHF) and through simultaneous saccharification and fermentation (SSF).

The SHF process is carried out in two independent reactors, each at its optimum pH and temperature conditions. While the SSF route allows it to carry out the enzymatic hydrolysis of the polysaccharides, together with the fermentation, using a single reactor. This last route offers operational advantages since it reduces the inhibition effects of the final product on enzymatic hydrolysis and immediate availability of fermentable sugars. However, the principal disadvantage is the need to find favorable conditions (temperature and pH) for enzymatic hydrolysis, fermentation, and the conflict to recycle the fermenting organism and enzymes. Overall, the SHF pathway is the best alternative in terms of intensifying the fermentation process, since the technoeconomic results show that the SSF pathway is a much more competitive process compared to the SHF pathway. The use of a single bioreactor results in a strong reduction in operating and investment costs (Faraco, 2013).

Regarding the yields, one of the highest reported in the literature was reached by Li et al. (2010). They achieved a bioethanol concentration of 22.3 g/L (84% of theoretical yield) through the integration of pentoses (xylan) and hexoses (glucose) by *E. coli* KO11 and *S. cerevisiae* D5A in the first and second phases, respectively. On the contrary, Karagöz et al. (2012), managed to obtain by *Saccharomyces cerevisiae* and *Pichia stipitis*, 14.07 g/100 g of raw straw and 5.73 g of bioethanol.

4.3.2.5 Recovery and purification of bioethanol

The principal problem in the production of anhydrous bioethanol (>99 mol.%) is the high energy cost involved in its separation. During the fermentation stage large quantities of fermentation broth are obtained with low concentrations of alcohol (between 5 and 12 wt.%) so it is necessary to eliminate excess water (Madson and Monceaux, 1995; Szitkai, et al., 2002).

Conventional methods for the recovery of anhydrous bioethanol from the fermentation broth contain at least three stages:

1. conventional distillation of dilute ethanol to a concentration close to its azeotropic point (95.57 wt.%);
2. extractive or azeotropic distillation using a third component to break up the azeotrope and remove the remaining water. Extractive distillation performs the separation in the presence of a relatively nonvolatile, high-boiling, miscible component that does not form an azeotrope with the other components of the mixture. For bioethanol—water extractive distillation, ethylene glycol continues to be the most widely used entrainment agent, although glycerol, hyperbranched polymers, and ionic liquids have also been proposed;
3. distillation to recover the third component and reuse it in the process.

The recovery of bioethanol by these methods implies the consumption of 50% to 80% of the total energy required in the entire bioethanol production process through fermentation. The recent interest in the search for clean and economic processes, and the strengthening of environmental legislation that restricts the use of solvents such as those used in azeotropic and extractive processes has led the industry to focus on other technologies such as extractive saline distillation, pervaporation and dehydration by molecular sieve adsorption (Szitkai et al., 2002).

However, these novel technologies are considerably expensive and have not been consolidated in the bioethanol-producing industry. Therefore, modifications to conventional distillation schemes that intensify the existing bioethanol purification process can result in economically attractive processes, and with their optimal dynamic behavior, sustainable processes can be achieved. This chapter will present a series of intensified schemes, thermally coupled distillation columns and divided wall columns (DWC), which are intended to be a palliative in the production costs of bioethanol.

4.3.3 Advantages and disadvantages of bioethanol production

The complexity and costs in the bioethanol production processes partly explain the reason why fuel bioethanol is not yet a replacement option for petroleum-derived fuels. However, there are a series of advantages to wagering in the obtainment of ethanol through biomass (Chiriboga Dechiara, 2014).

- They suppress renewable resources and significantly reduce by partial or total substitution the amounts used by fossil fuels. This leads to a decrease in the use and import of hydrocarbons and favors the use of renewable natural resources. It can counteract and reduce the impact of recurring hydrocarbon price and reserve problems in periods of energy crisis.
- It promotes economic and productive reactivation. It establishes an agrochain where public and private links are integrated, such as the sequence of the agricultural, industrial, energy, social, economic, and environmental sectors. These sectors, primarily agricultural and industrial, would create direct and indirect sources of work, promoting rural employment, industrial regionalization, and the development of regions with agroindustrial potential.
- It has superior physical properties. It has a high degree of solubility and miscibility with gasoline. Bioethanol, being a liquid is suitable for use in automobiles. It has a high octane number and does not affect the quality of gasoline for use in combustion engines.
- Its use as fuel favors the reduction of atmospheric pollution, especially in urban centers, due to the lower production of carbonaceous particles. It reduces the emission of CO_2 compared to gasoline and it contributes to the improvement of health and quality of life.

Unquestionably, there are disadvantages in the production and use of bioethanol as an energy source. The most important ones being:

- Its production and price will be directly tied to the national and international market of the raw material in question.
- Ethanol has 30% less energy density compared to gasoline, which requires higher fuel consumption.
- On the other hand, the high volatility of anhydrous ethanol affects the efficiency of the mixture and can lead to intermittences in the supply of fuel to the engine and losses due to evaporation.
- It has special infrastructure requirements for transportation and storage. Structurally and mechanically adapted vehicles are needed to run on bioethanol fuel.
- The agricultural production of biomass requires large areas of land, which could otherwise be used for crops.

However, and despite these disadvantages, two central problems have made bioethanol a real and viable alternative in the energy market. These are conflicts concerning the environment and the periodic conflicts that arise from countries that export oil. More recently, the development of effective technologies in the production of bioethanol has become a priority for several research centers and universities as well as for numerous governments that have invested in this area.

4.4 Problems of the process for obtaining conventional bioethanol

To be competitive, and find economic acceptance, the cost for bioconversion of biomass into liquid fuel must be lower than the current gasoline prices (Wayman, 1990; Subramanian et al., 2005); however, there is still a wide margin to reduce the cost of converting biomass in bioethanol.

The cost of raw material and cellulolytic enzymes are two important parameters in the production of low-cost bioethanol. Raw material represents around 40% of the cost of bioethanol production (Hamelinck et al., 2005). In the same way, coupled with the high cost of the raw material, there is energy expenditure in the separation of bioethanol. The selection of the separation method is intrinsically linked to the type of raw material. In recent years, investigations have been presented that dictate that the use of process intensification leads to significant savings in energy expenditure and equipment cost. Implementing the intensification in the bioethanol purification processes through thermally coupled distillation systems and DWC schemes the principal discussion about the economy of liquid biofuels is mitigated. This is part of the resource allocation between the energy and agriculture sectors since in many places the costs to produce crops and convert them into bioethanol is too high to compete in the market against fossil fuels (FAO, 2008).

This chapter offers one of the most important tools to reduce the cost of bioethanol production, such as the intensification of the purification process to reduce energy demand and aim to achieve a sustainable process with optimum system control. It should be noted that the intensification of processes consists of the development of new equipment and techniques that, compared to those commonly used today, are expected to bring improvements in the manufacture and processing of bioethanol, which substantially reduce the size/production, energy consumption or waste production, and ultimately results in cheaper processes and sustainable technologies.

4.5 Proposals to intensify the process for obtaining bioethanol

The separation of the bioethanol–water mixture by conventional distillation is limited by the presence of an azeotrope. To obtain high purity bioethanol by distillation, certain techniques have been developed to alter the relative volatilities of the substances in the mixture and thereby allow the azeotrope to break down. Among these techniques, the most commonly used are vacuum distillation, azeotropic distillation, and extractive distillation. In this chapter, designs and operating conditions that maximize the utility function for the extractive distillation of fuel grade bioethanol using glycerol or ethylene glycol as the entraining agent are proposed. This aims to offer the greatest economic benefit in stationary conditions. However, to maintain the optimal operating conditions of the process, a control study of said systems was carried out, so that in practice, when processes are subjected to disturbances and/or process transitions, they result in sustainable processes.

The importance of considering sustainability issues in the early stages of intensified process design can help differentiate between processes that are easy and processes that are difficult to operate. According to Jiménez-González et al. (2012) "green metrics" should be incorporated when designing a process toward the broader objective of environmental sustainability. Among those green metrics, the economic and process control aspects should be highlighted. In addition, the 11 principles of green chemistry express the desire for real-time process monitoring and analysis. The goal of this principle is to avoid waste and safety issues (reflected in control properties) by identifying process deviations as they occur. In other words, energy-efficient separation processes for bioethanol purification are today more essential than ever to speed up the use of biofuels. It is for this reason that both energy consumption and capital costs will be evaluated, and thus be able to identify some promising intensified distillation configurations for the purification stage in the production of fuel bioethanol. The process control analysis finalizes the generation of synthesis alternatives and in many cases, it is essential to highlight some design drawbacks. The evaluation of these metrics will provide a broad view of how different variables can affect the sustainability of the process.

4.5.1 Synthesis

Three transcendental stages will be identified during the "bioethanol separation sequence design procedure." These stages include the definition of a methodology to generate the intensified sequences, the definition of what type of configurations should be included in the search space (simple columns or thermally coupled columns) and the analysis of control properties.

To carry out the synthesis of the intensified sequences, the methodology described by Errico and Rong (2012) was used. Errico and Rong (2012) present a methodology to synthesize the modified extractive distillation configurations for the purification of bioethanol. The systematic method of

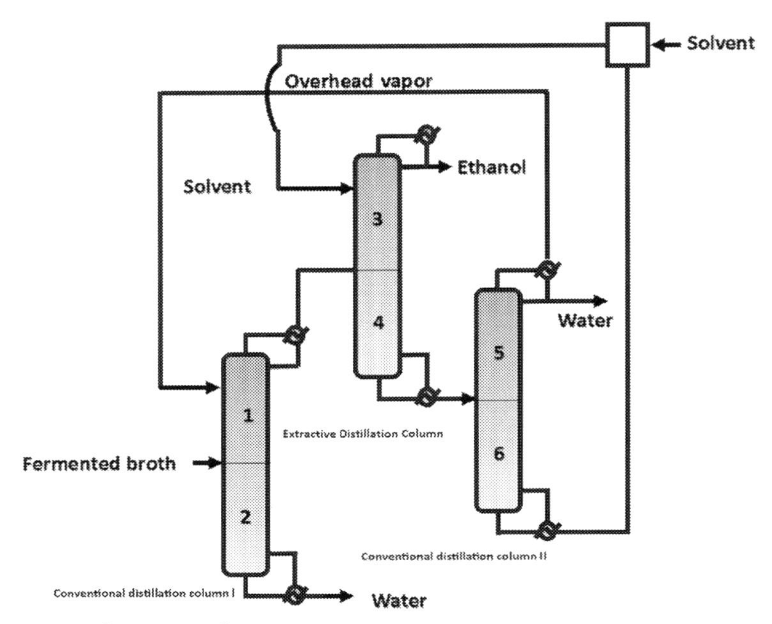

Figure 4.4 Reference configuration.

Errico and Rong (2012) shows four stages. Each stage will present the specific modifications envisioned to the reference configuration (Fig. 4.4).

- Stage 1: identification of the corresponding simple column.

  The reference sequence shown in Fig. 4.4 is used. In the first column, the heavier feed component is separated, while in the second column, the lighter feed is obtained as distillate. In the third column, the remaining components are recovered. Each section of the column was designated with the number as can be seen in the figure. A column section is traditionally defined as a portion of the distillation column that is not interrupted by the entry or exit of streams or heat flows. Since recyclable streams exist, they modify sections 1 and 3 of the reference setup.
- Stage 2: Definition of the sequences thermodynamically.

With the reference sequence sectioned, it is possible to generate a series of sequences thermodynamically by replacing one or more heat exchangers associated with nonproduct streams with a liquid–vapor interconnection. Even if a subspace does appear in the distillation sequences for the generation of intensified schemes, it is not entirely promising in terms of improving the separation efficiency for the production of bioethanol, and this is a

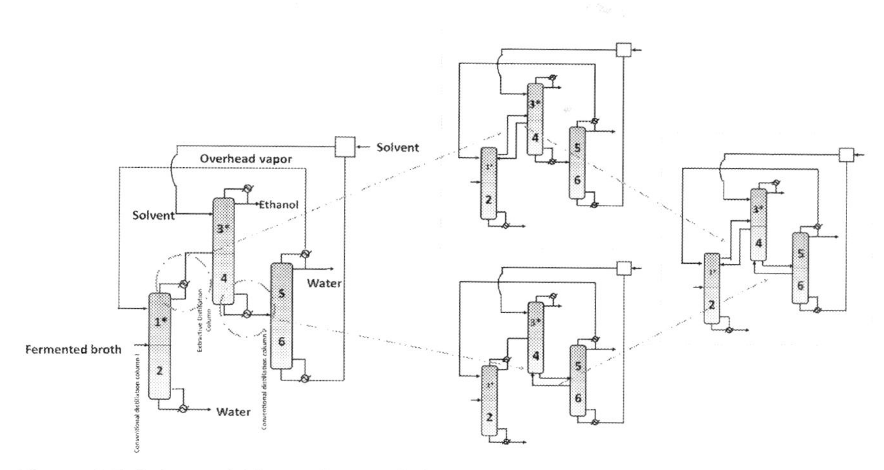

Figure 4.5 Enhanced (thermally coupled) settings for extractive distillation.

fundamental step of the design procedure to achieve the final configurations in the next stages. The configurations shown in Fig. 4.5 show the replacement of the condenser and/or reboiler by a bidirectional connection of liquid and vapor.

In the case of extractive distillation, the presence of the solvent stream and the recyclables could give a preliminary indication of the applicability of a particular configuration. Assuming the liquid stream from the second to the first column is essentially made up of the high-boiling solvent. This restricts the maximum purity obtainable for the water stream, so no positive improvements should be expected from these settings.

- Stage 3: Definition of thermodynamically equivalent sequences.

By having ideal mixtures, for thermally coupled columns it is possible to create corresponding equivalent thermodynamic configurations by moving a column section that provides the common reflux ratio or boiling vapor between two consecutive columns. Fig. 4.6 shows the methodology of section movements and a possible recombination of the thermally coupled configuration.

- Stage 4: Generation of lateral flow sequences.

Single column sections of thermodynamically equivalent sequences can be supplemented with lateral liquid or vapor extraction or even thermal coupling. In some cases, this turns out to be profitable (depending on the feed composition). By applying this procedure to the azeotropic mixture, it is possible to generate some configurations that are reported in the following design section.

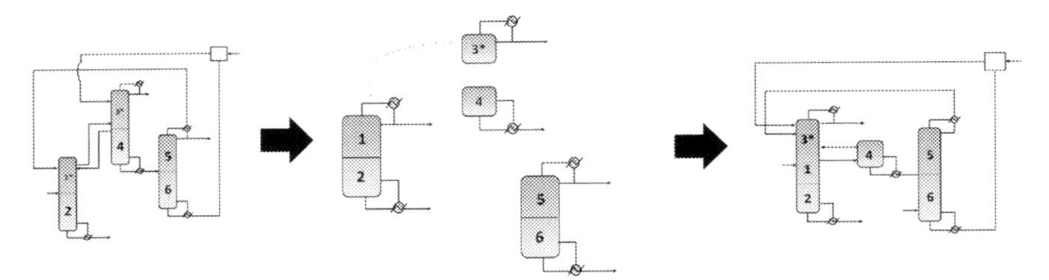

Figure 4.6 Modified thermodynamically equivalent configuration for extractive distillation.

4.5.2 Design

Under the synthesis methodology shown above, it was possible to obtain a series of designs that claim to be economically superior to conventional bioethanol purification processes and with sustainability benefits. The designs obtained are the following:

- SDC (Sequence Dual Columns): this configuration consists of a prefractionator followed by the extractive distillation column with a side stream. In the extractive column, bioethanol is obtained in the upper part, the solvent (glycerol-SDC or ethylene glycol-SDC-A) is recovered in the lower part and the side stream of steam contains a mixture of water and bioethanol, which is recycled to the prefractionator (see Fig. 4.7A).
- STCL (Sequence Three-Column separation with Liquid): this configuration consists of a prefractionator, in which the water–bioethanol mixture is partially separated until a purity close to the azeotrope is reached, followed by an extractive distillation column where, using an extractive agent (ethylene glycol), almost pure bioethanol is obtained in the upper section of the column. Finally, the solvent is recovered in the last column of the sequence. In the upper section of the solvent recovery column, a liquid water–ethanol mixture is obtained, which is recycled to the prefractionator (see Fig. 4.7B).
- STCV (Sequence Three-Column separation with Vapor): This bioethanol–water mixture configuration is obtained in the upper part of the solvent recovery column in the vapor phase (STCV). The extractive agents used are glycerol (STCV) or ethylene glycol (STCV-A) and can be seen in Fig. 4.7C.
- DWCL (Dividing Wall Column with Liquid) sequence: A configuration in which a liquid stream containing water and bioethanol is recycled to the first column; ethanol is obtained as a distillate from the main column, while solvent (ethylene glycol) is recovered in the bottoms (see Fig. 4.7D).

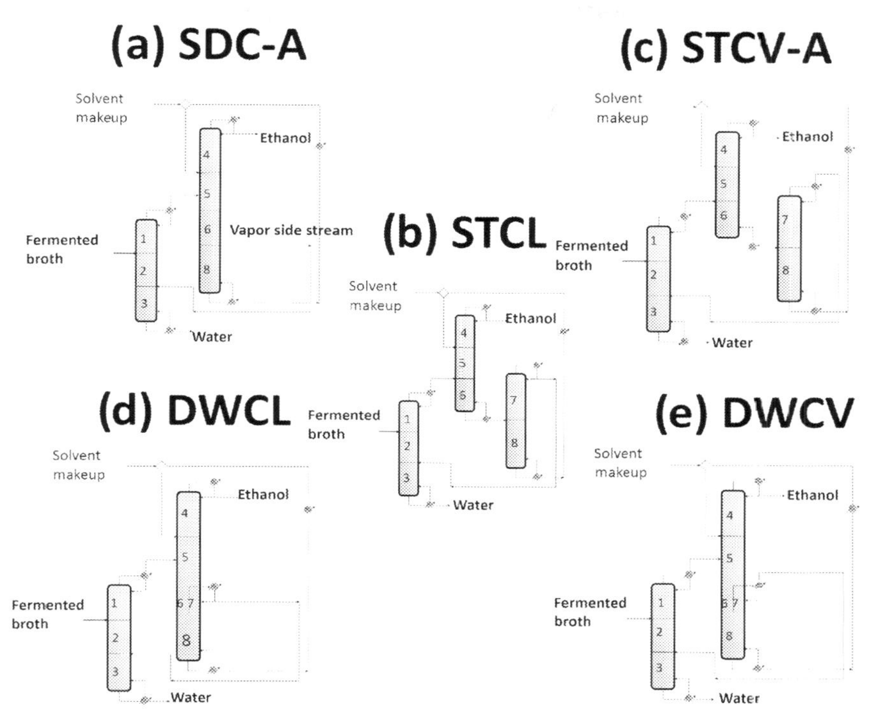

Figure 4.7 Proposed thermodynamically equivalent configurations.

- DWCV (Dividing Wall Column with Vapor) sequence: A configuration similar to the previous one, but the recycled ethanol—water mixture is obtained in the vapor phase (see Fig. 4.7E).

To compare the performance of the different alternative configurations, a feed stream of 1,694,240 kmol/hour of a bioethanol—water mixture containing 5% mol of bioethanol was considered. The minimum concentration of bioethanol was defined as 0.999 based on the mole fraction, and the NRTL (NonRandom Two-Liquid) method was used to evaluate the activity coefficients. All the sequences considered were simulated using the Aspen Plus V8.8. For all simulations, an optimal value of 0.87 was defined as a solvent to feed ratio. When necessary, a sensitivity analysis was used to optimize parameters and meet purity goals. The cost of equipment was carried out using the Aspen Plus Economic Analyzer. Capital costs were annualized considering an average operating time of 10 years. The Aspen Plus economic evaluator is not based on simple module factors. However, if extensive data is used to estimate material, labor and

Table 4.3 Energy requirements, capital costs, and solvent composition for the configurations considered using ethylene glycol as solvent.

	Total condenser duty (kW)	Total reboiler duty (kW)	Annualized capital cost (k$/year)	Solvent makeup (kmol/h)
SDC-A	3852.16	4902.93	107.8	0.62
STCL	4081.32	5118.425	132.8	0.004
STCV-A	3871.04	4907.89	133.1	0.004
DWCL	3970.82	5005.80	102.3	0.004
DWCV	3800.72	4837.32	102.6	0.004

Table 4.4 Energy requirements, capital costs, and solvent composition for configurations using glycerol as the solvent.

	Total condenser duty (kW)	Total reboiler duty (kW)	Annualized capital cost (k$/year)	Solvent makeup (kmol/h)
SDC	3750.638	4934.136	112.9	0.010
STCV	3820.65	5018.67	134.3	0.001

construction costs, Table 4.3 shows energy consumptions, capital costs, and solvent feed rate for configurations using ethylene glycol.

It should be noted that the DWCL and DWCV configurations are promising options that reduce energy and capital costs. However, considering the complexity of its structure, a dynamic analysis is essential. Another interesting configuration is the SDC-A. It is evident that the SDC-A configuration has a high solvent consumption due to the presence of the side stream. Using this type of setup can be convenient as it can be assessed as a difference between energy and capital costs saved compared to traditional sequences and increased solvent consumption. For this reason, the STCV and SDC configurations were simulated considering glycerol as solvent. Glycerol is a cheaper solvent that can be obtained as a by-product from biodiesel plants. The results when glycerol is used as the entraining agent are shown in Table 4.4.

Although the selected configurations are shown as the best options for the production of bioethanol, they are reconsidered here for the analysis of their dynamic behavior. Table 4.5 shows the design parameters of the best configurations. The study of control properties represents the last step in selecting the most promising green metrics configuration toward the broader goal of environmental sustainability.

Table 4.5 Design parameters of the best configurations.

	SDC		SDC-A		STCV-A			DWCL		DWCV	
	Col I	Col II	Col I	Col II	Col I	Col II	Col III	Col I	Col II	Col I	Col II
Stages	44	36	44	42	44	29	9	45	43	45	43
Reflux ratio (mol)	2.338	0.198	2.33	0.42	2.39	0.15	0.580	2.54	0.32	0.33	2.33
Feed stage	29	25	30	25	30	26	3	30	26	30	26
Solvent feed stage	—	3	—	4	—	2	—	—	4	—	4
Diameter (m)	1.365	0.80	1.4	0.85	1.39	0.75	0.65	1.36	0.80	1.33	0.80
Vapor side stream stage	—	28	—	28	—	—	—	—	—	—	—

4.5.3 Control

Analyzing the dynamic behavior of the alternatives shown helps to define their applicability. The process control analysis completes the generation of alternatives and synthesis. In many cases, it is essential to highlight some design drawbacks. To carry out the dynamic simulation in closed-loop control of the extractive distillation column, a certain methodology has to be determined for the derivation of the optimal sequence in terms of control, several phases of simulation and subsequent analysis of the results.

The tuning of the Proportional-Integral (PI) controllers is worked on the dynamic files of Aspen Dynamics. The loops to tune are those indicated by the heuristic on the loop closest to our variable in question (see Fig. 4.8). For this study, the purity of the bioethanol was adjusted with reflux in all sequences and the purity of the water varied depending on the sequence, either with reflux or with the flow of the side stream. The algorithm is presented in Fig. 4.9.

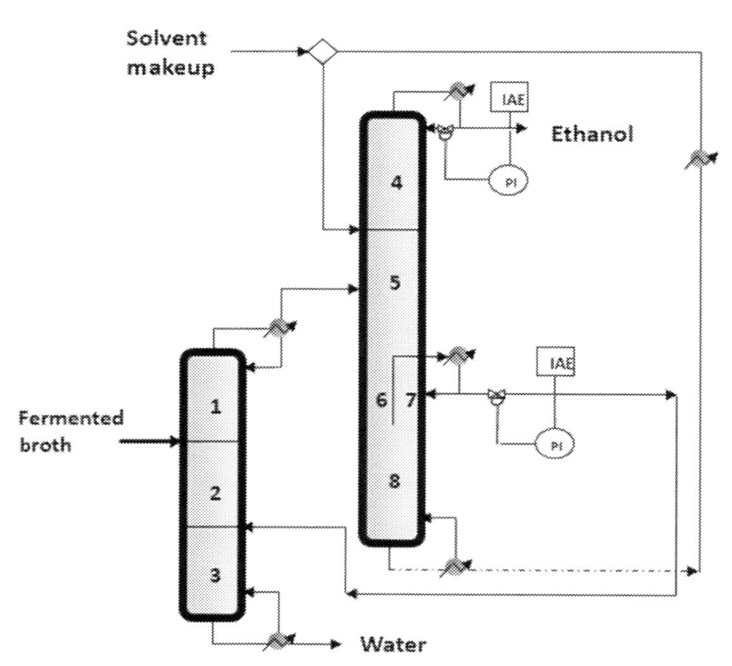

Figure 4.8 Control loops in the DWCL configuration.

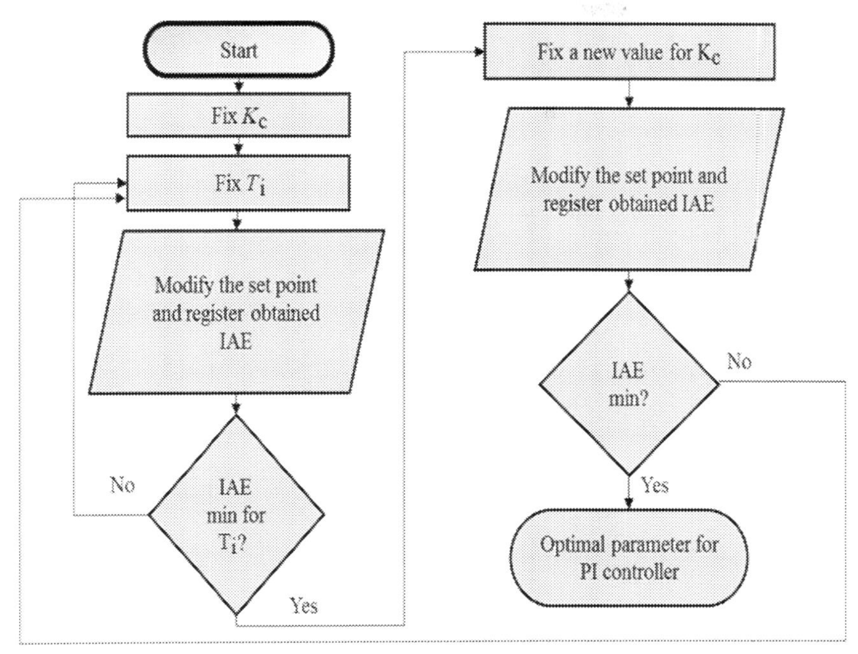

Figure 4.9 IAE (Integral of the Absolute value of the Error) minimization methodology.

Table 4.6 Closed-loop tuning results.

	Ethanol			Water		
	K_c	T_i	IAE	K_c	T_i	IAE
SDC	250	25.4	0.008570903	25	3	0.000490910
SDC-A	20	68.00	0.014097702	175	8.75	0.002797684
STCV	250	22.76	0.008675855	—	—	—
STCV-A	250	55	0.010519718	225	61.5	0.012179141
STCL	250	51.25	0.010172300	150	100	0.015901120
DWCV	225	58.25	0.010521612	10	139.75	0.050686177
DWCL	250	52	0.011971880	—	—	—

With this methodology, the results are presented in Table 4.6. It can be inferred that the best sequence is the SDC since it results with the lowest IAE value for the bioethanol component and the second best value for water; the best response in the water loop is for the SDC-A sequence, but this has the highest IAE value for bioethanol.

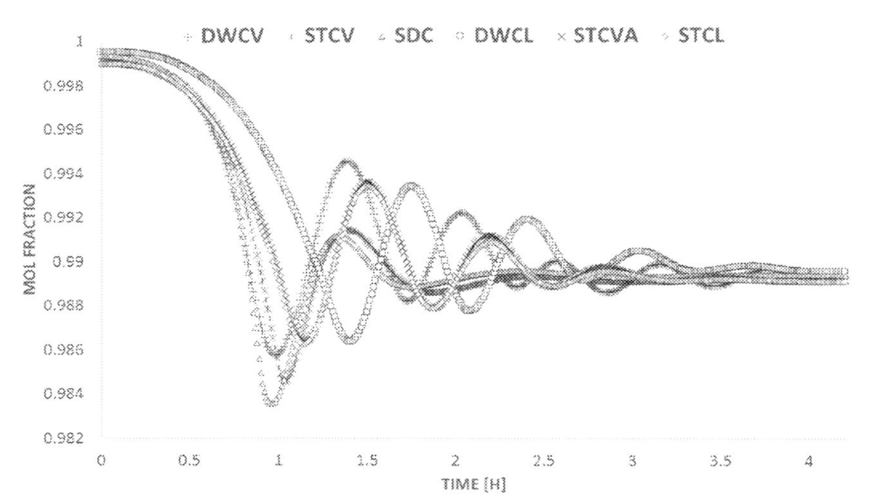

Figure 4.10 Behavior of the composition with respect to time (bioethanol).

It is also important to observe the dynamic behavior that is generated in the simulation in the form of a graph, which dictates how the composition of the element under analysis varies with respect to time. The oscillation present in these graphs affects the search for the sequence with the best controllability response, in conjunction with the IAE indicator. In Fig. 4.10, it can be seen that there is congruence with what is shown by the indicator, the oscillation that shows a faster stabilization despite being the one that presents a greater overdraft. It is clear that the response, in the bioethanol tie, is much faster than for the other sequences. Now for the water loop, Fig. 4.11 shows that the response of the SDC sequence is much faster with respect to the others.

It can be affirmed that for the heuristic method the best sequence is the sequence of two columns with glycerol solvent (SDC). To corroborate the reliability of the SDC sequence, a study was submitted in which the system was continuously disturbed on several occasions waiting for it to respond and achieve stabilization. The disturbances that were carried out in the system were the following: first, a negative percentage of 1% of the original composition of the output current was disturbed. Second, without restoring the system to its initial conditions it was returned to the original composition. Third, the composition of the original outlet current was disturbed to a negative 1.5% and returned to the original composition. With this study, it was observed that the SDC sequence is capable of stabilizing the system to different perturbations in the compositions (see

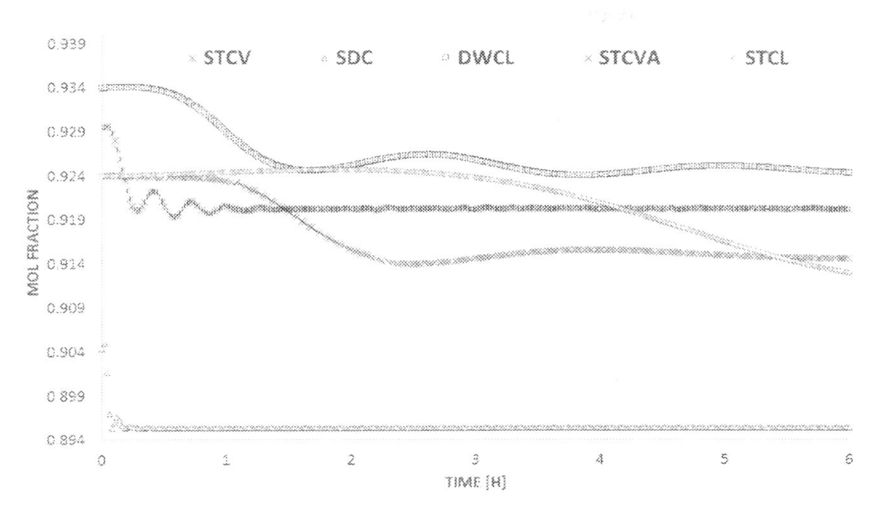

Figure 4.11 Behavior of the composition with respect to time (water).

Table 4.7 Simulation results with continuous disturbances.

	K_c	Ti	IAE 1%	IAE	IAE 1.5%	IAE
SDC	250	25.4	0.008580998	0.019642446	0.031807626	0.042052723
	25	3	4.90625E-04	1.40819E-03	2.95452E−03	0.004501014

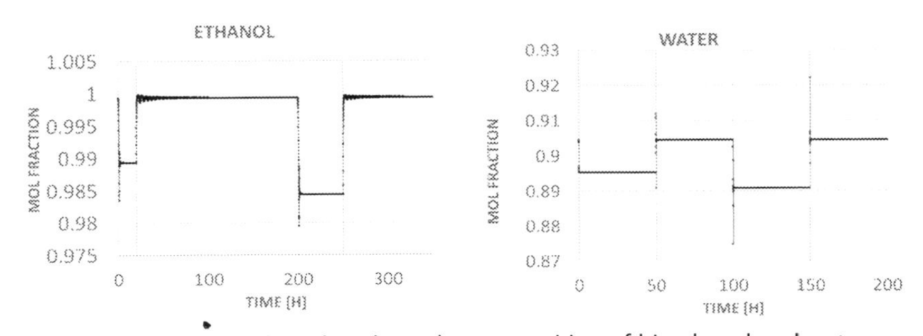

Figure 4.12 Behavior when disturbing the composition of bioethanol and water.

Table 4.7), this reaffirms the reliability of the sequence in terms of control is the SDC (see Fig. 4.12).

With this, this chapter offers important alternatives to reduce the cost of bioethanol production and reduce energy demand by trying to achieve a sustainable process with optimal control of the system.

4.6 Conclusions

The increase in bioethanol production in the world has been offset by the development of new technologies that make it possible to obtain bioethanol from wood waste, solid waste, and all materials containing cellulose and hemicellulose, which makes it possible to revalue waste of various industries to turn them into raw material to obtain bioethanol. However, new alternatives must be sought after to purify bioethanol through low-cost and sustainable separation processes in order to maintain competitive prices compared to those of traditional fossil fuels. Aside from showing the basic aspects in the obtainment of bioethanol, this chapter shows real alternatives of intensified processes to be able to mitigate the economic and energy cost and seeks the best control properties to give a character of sustainability to the process. The results of the intensified sequences, both in the technoeconomic area and in the closed-loop control indicate that, overall, the configuration with a side stream of steam, and using glycerol as solvent (SDC) is the best option. Configurations with a side stream and conventional sequences using glycerol as a solvent outperformed the dynamic behavior of the dividing wall schemes. In addition to this, the results also suggest that the economic aspect and the control responses are governed by the type of solvent used. The better performance of the sequences with glycerol compared to ethylene glycol further promotes the use of sustainable solvents and this helps to generate sustainable and environmentally friendly processes. The possibility of generating, with the synthesis methodology shown, new configurations that aid in obtaining bioethanol, positively contributing to the biofuels category, are also welcomed.

References

Alternative Fuels Data Center, 2020. World ethanol production, United States DOE. <http://www.afdc.energy.gov/> (last accessed 05.07.2020).

Alvira, P., Tomás-Pejó, E., Ballesteros, M., Negro, M.J., 2010. Pretreatment technologies for an efficient bioethanol production process based on enzymatic hydrolysis: a review. Bioresour. Technol. 101 (13), 4851–4861.

Balat, M., 2011. Production of bioethanol from lignocellulosic materials via the biochemical pathway: a review. Energy Convers. Manag. 52 (2), 858–875.

Chiriboga Dechiara, P.A. (2014). La producción de biocombustibles frente al uso del territorio para garantizar la soberanía alimentaria y el régimen del buen vivir. Mecanismos legales para desarrollar los biocombustibles en base al derecho comparando, Bachelor'sthesis, Universidad de las Américas, Quito.

Demirbaş, A., 2005. Bioethanol from cellulosic materials: a renewable motor fuel from biomass. Energy Sources 27 (4), 327–337.

Errico, M., Rong, B.G., 2012. Synthesis of new separation processes for bioethanol production by extractive distillation. Sep. Purif. Technol. 96, 58–67.

Evans, M.K., 1997. The Economic Impact of the Demand for Ethanol. DIANE Publishing.

FAO (Food and Agriculture Organization of the United Nations), (2008). *The state of food and agriculture. Biofuels: perspectives, risks and opportunities.* Viale delle Terme di Caracalla 00153 Rome Italy. <http://www.fao.org.> ISSN 0251–1371. (last accessed 12.08. 2020).

Faraco, V. (Ed.), 2013. Lignocellulose Conversion: Enzymatic and Microbial Tools for Bioethanol Production. Springer Science & Business Media, pp. 131–135.

Fougret, C.M., Hölderich, W.F., 2001. Ethylene hydration over metal phosphates impregnated with phosphoric acid. Appl. Catal. A: Gen. 207 (1–2), 295–301.

Gómez Tovar, F. (2008). Métodos secuenciales de pretratamiento químico y enzimático de residuos agrícolas para la producción de metano, *Master's thesis*, pp. 87–96.

GVR Grand View Research, (2020). *Ethanol market size, share & trends analysis report by source (second generation, grain-based), by purity (denatured, undenatured), by application (beverages, fuel & fuel additives), and segment forecasts, 2020–2027, United States.* <https://www.grandviewresearch.com/industry-analysis/ethanolmarket#:~:text=Report%20Overview,the %20product%20as%20a%20biofuel> (last accessed 20.07.2020).

Hamelinck, C.N., Van Hooijdonk, G., Faaij, A.P., 2005. Ethanol from lignocellulosic biomass: techno-economic performance in short-, middle-and long-term. Biomass Bioenergy 28 (4), 384–410.

Hernández-Nodarse, M.T., (2007). *Current trends in bioethanol production, Lecture given at the Faculty of Engineering—TEC Landívar*, University Rafael Landívar, 1–17.

Hidzir, N.S., Som, A.S., & Abdullah, Z. (2014). Ethanol production via direct hydration of ethylene: a review. In *International conference on global sustainability and chemical engineering* (ICGSE).

Jeong, J.S., Jeon, H., Ko, K.M., Chung, B., Choi, G.W., 2012. Production of anhydrous ethanol using various PSA (pressure swing adsorption) processes in pilot plant. Renew. Energy 42, 41–45.

Jiménez-González, C., Constable, D.J., Ponder, C.S., 2012. Evaluating the "Greenness" of chemical processes and products in the pharmaceutical industry—a green metrics primer. Chem. Soc. Rev. 41 (4), 1485–1498.

Karagöz, P., Rocha, I.V., Özkan, M., Angelidaki, I., 2012. Alkaline peroxide pretreatment of rapeseed straw for enhancing bioethanol production by same vessel saccharification and co-fermentation. Bioresour. Technol. 104, 349–357.

Li, X., Kim, T.H., Nghiem, N.P., 2010. Bioethanol production from corn stover using aqueous ammonia pretreatment and two-phase simultaneous saccharification and fermentation (TPSSF). Bioresour. Technol. 101 (15), 5910–5916.

Loïc, A. B. B. A. D., Daryl, F. R. F. A. H., Sacha, L. T. P. J. P., & Julien, R. A. P. P. A. Z. Z. O, 2019. Ethanol Production by catalytic hydration of ethylene, 1–21.

Madson, Py, Monceaux, D., 1995. Fuel ethanol production. In: Lyons, T.P., Kelsall, D.R., Murtagh, J.E. (Eds.), The Alcohol Textbook. University Press, Nottingham, pp. 257–268.

Mosier, N., Wyman, C., Dale, B., Elander, R., Lee, Y.Y., Holtzapple, M., et al., 2005. Features of promising technologies for pretreatment of lignocellulosic biomass. Bioresour. Technol. 96 (6), 673–686.

Mussatto, S.I., Roberto, I.C., 2004. Alternatives for detoxification of diluted-acid lignocellulosic hydrolyzates for use in fermentative processes: a review. Bioresour. Technol. 93 (1), 1–10.

Oliva Domínguez, J.M. (2004). Efecto de los productos de degradación originados en la explosión por vapor de biomasa de chopo sobre "Kluyveromyces marxianus", Doctoral dissertation, Universidad Complutense de Madrid, Servicio de Publicaciones.

Ovando-Chacón, S.L., Waliszewski, K.N., 2005. Preparativos de celulasas comerciales y aplicaciones en procesos extractivos. Univ. y. Cienc. 21 (42), 113–122.

Qi, H., Zhou, X.G., Liu, L.H., Yuan, W.K., 1999. A hybrid neural network – first principles model for fixed-bed reactor. Chem. Eng. Sci. 54 (13–14), 2521–2526.

Rutz, D., & Janssen, R. (2007). Biofuel technology handbook. WIP Renew. Energ., 95.

Sánchez Riaño, A.M., Gutiérrez Morales, A.I., Muñoz Hernández, J.A., Rivera Barrero, C.A., 2010. Bioethanol production from agroindustrial lignocellulosic byproducts. Rev. Tumbaga 5, 61–91.

Sanchez, O.J., Cardona, C.A., 2008. Trends in biotechnological production of fuel ethanol from different feedstocks. Bioresour. Technol. 99 (13), 5270–5295.

Sindhu, R., Binod, P., Pandey, A., Ankaram, S., Duan, Y., & Awasthi, M.K., 2019. Biofuel production from biomass: toward sustainable development. In: *Current Developments in Biotechnology and Bioengineering*. Elsevier, 79–92.

Subramanian, K.A., Singal, S.K., Saxena, M., Singhal, S., 2005. Utilization of liquid biofuels in automotive diesel engines: an Indian perspective. Biomass Bioenergy 29 (1), 65–72.

Szitkai, Z., Lelkes, Z., Fonyo, Z., 2002. Optimization of hybrid ethanol dehydration systems. Chem. Eng. Process: Process. Intensif. 41 (7), 631–646.

Talebnia, F., Karakashev, D., Angelidaki, I., 2010. Production of bioethanol from wheat straw: an overview on pretreatment, hydrolysis and fermentation. Bioresour. Technol. 101 (13), 4744–4753.

Wayman, M., 1990. Biotechnology of biomass conversion. In: Wayman, M.T., Parekh, S.R. (Eds.), Fuels and chemicals from renewable resources. Open University Press, Milton Keynes, pp. 1586–1597.

Weissermel, K., Arpe, H.J., 2008. Industrial Organic Chemistry. John Wiley & Sons, p. 203.

Yamamoto, H., 2017. An analysis of supply and demand strategy of bioethanol using an agent-based global energy model. J. Jpn. Inst. Energy 96 (7), 239–249.

CHAPTER 5

Biobutanol

Contents

5.1 General characteristics, uses, and applications

Butanol and its isomers have a four-carbon structure, either a linear structure or a branched structure. The variation in the structural topology of the molecules, as well as the position of the functional group OH, produces variations in alcohol properties. Table 5.1 provides a general description of the physicochemical properties of butanol and its isomers (Jin et al., 2011).

According to Table 5.1, the applications of the different isomers are similar. Additionally, all isomers can be produced from oil and biomass. In particular, n-butanol has the best characteristics for use as a liquid fuel or as an additive to other fuels. Some of the main characteristics of gasoline, diesel, ethanol, and n-butanol are found in Table 5.2 (Bankar et al., 2013; Jin et al., 2011).

Analyzing specific data present in Table 5.2, it is possible to identify some interesting characteristics that place butanol as a compound with the potential to overcome some disadvantages present in petroleum-based fuels.

Improvements in Bio-Based Building Blocks Production Through Process Intensification and Sustainability Concepts
DOI: https://doi.org/10.1016/B978-0-323-89870-6.00002-0

Table 5.1 Butanol isomers, topology, and uses.

Isomer	Molecular topology	Main uses
1-Butanol (n-butanol)	OH	Solvent (painting), plasticizers, hydraulic brake fluid, cosmetics, gasoline additive
2-Butanol	OH	Solvent, domestic cleaning agent, paint remover
iso-Butanol	OH	Solvent and additive, remove ink ingredient, gasoline additive
tert-Butanol	OH	Solvent, industrial cleaner, intermediate for metil tert-butyl ether (MTBE), ethyl tert-butyl ether (ETBE)

For example, butanol presents:

1. Higher heating value. Normally, a higher heating value is associated with higher carbon content. In this sense, butanol has a 50% higher energy density by volume. Lower volatility. Contrary to the heating value, the volatility decreases when the carbon content increases. The decrease in volatility is related to a lower tendency to be vaporized, which consequently indicates that butanol is potentially safer.
2. Intersolubility. Commonly, the compounds with a higher number of carbon present greater ease to be mixed with gasoline. In contrast to some shorter chain compounds that present polarity characteristics, being potentially more soluble with water.
3. Higher viscosity. As a consequence of having a higher viscosity, butanol has the potential to generate protection in some engine components that are in direct contact with the fuel.

On the other hand, butanol also has some limitations and disadvantages when used as fuel in spark-ignition engines (Merola et al., 2012; Rakopoulos et al., 2010). Some of these are listed below.

Table 5.2 Properties of conventional fossil fuels and some alcohols.

	Gasoline	Diesel	Methanol	Ethanol	n-Butanol
Molecular formula	C_4-C_{12}	$C_{12}-C_{25}$	CH_3OH	C_2H_5OH	C_4H_9OH
• Molecular weight	111.19	198.4	32.04	46.06	74.11
• Octane number	80–99	20–30	3	8	25
• Oxygen content (%weight)	–	–	50	34.8	21.6
• Density (g mL^{-1}) at 20°C	0.72–0.78	0.82–0.86	0.796	0.79	0.808
• Autoignition temperature (°C)	300	210	470	434	385
• Lower heating value (MJ kg^{-1})	42.7	42.5	19.9	26.8	33.1
• Boiling point (°C)	25–215	180–370	64.5	78.4	117.7
• Flammability limits (%vol)	0.6–8	1.5–7.6	6–36.5	4.3–19	1.4–11.2
• Saturation pressure (kPa) at 38°C	31.01	1.86	31.69	13.8	2.27
• Viscosity (mm^2 s^{-1}) at 40°C	0.4–0.8	1.9–4.1	0.59	1.08	2.63
• Energy density (MJ L^{-1})	32	35.86	16	19.6	29.2

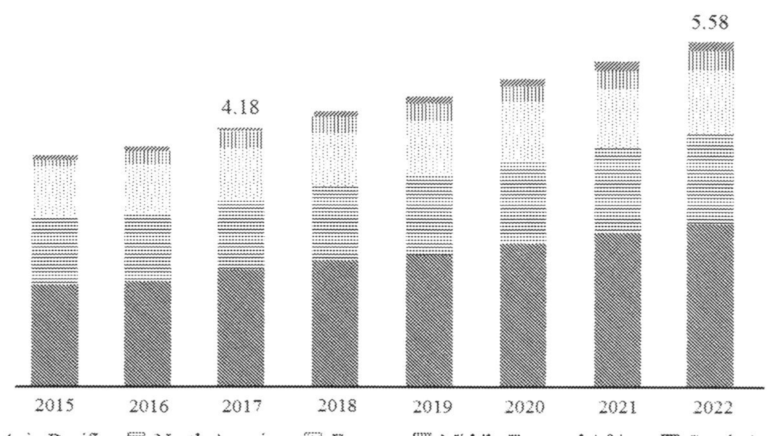

Figure 5.1 Global demand of butanol by geographic region.

1. Lower engine performance. Since butanol has a lower heating value, operating conditions can be generated where the engine operates with lower performance compared to what it would have if it operated with gasoline.
2. Higher fuel consumption. Due to the lower heating value compared to the use of gasoline, the engine will demand a greater amount of butanol to be injected into the combustion cycle.
3. Since butanol has a lower octane rating than ethanol, it is recommended for use in engines with lower compression ratios, resulting in lower performance.
4. N-butanol has fewer cetanes compared to diesel or biodiesel, reducing self-ignition and potentially impairing combustion control.

As far as global demand for butanol is concerned, there are records of steady growth since 2017. In 2017, sales were estimated at 4180 million dollars and are expected to reach 5580 million dollars in 2022, that is, a growth of 5.9%. The high demand in various regions is due to the versatility of butanol, either as a product or as a precursor to other products. Fig. 5.1 shows sales associated with butanol in various regions of the world.

5.2 Production of butanol from fossil sources

Traditionally, butanol is produced from fossil fuels, better known as petro-butanol. Currently, several corporations produce synthetic and

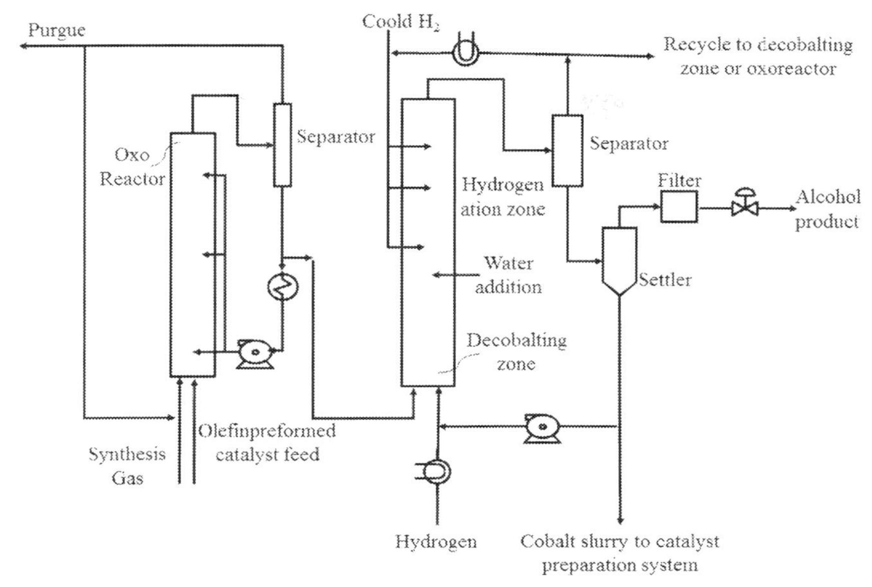

Figure 5.2 Butanol production through the oxo process.

petroleum-derived preoil. For example, BASF SE (Germany), DOW (United States), BASF-YPC (China), OXO Corporations (United States), Formosa (Taiwan), Oxichimie (France), to mention a few industries, currently produce butanol (Mariano et al., 2013).

Synthetically produced butanol is obtained from the hydroformylation of propylene, better known as the oxo process (observe Fig. 5.2). In that sense, butanol production by this route is strongly linked to the price of propylene and evidently to the price of oil (Dürre, 2011).

The oxo process is the reaction of carbon monoxide and hydrogen with olefin as a substrate to form isomeric aldehydes as shown in Eq. (5.1). The ratio of isomeric aldehydes depends on the olefin, the catalyst and the reaction conditions.

$$RCH = CH_2 + CO + H_2 \xrightarrow{\text{Catalyst}} RCH_2CH_2CHO + R(CH_3)CHCHO \tag{5.1}$$

The oxo reaction often proceeds in the presence of metals of the 8–10 group in liquid phase. In the first procedures of the oxo process, catalysts based on cobalt were used, in a process that demanded high pressure and high contents of hydrogen and carbon monoxide, generating a somewhat unstable process and in the presence of toxic substances. However, because the oxo process is a direct way of converting cheap olefins into high value-added building blocks, research into this process increased significantly in Europe, Japan, and the United States.

The challenges in the oxo process are to simultaneously achieve high production rates, as well as high selectivity for the desired compound, and to use a stable catalyst. In addition, progress has been made in the development of high reactivity rhodium catalysts for the conversion of internal and mixed-olefin feed streams. These latter are considerably less reactive than simple unsubstituted α-olefins. Development of catalysts that give improved process selectivities to the straight-chain isomer, generally more valuable, and of more efficient ways to recover product from rhodium catalyst solutions, have occurred.

5.3 Butanol production by the biochemical route

The most attractive and theoretically sustainable route is through the fermentation of sugar (Chen et al., 2014), glycerol (Yadav et al., 2014) or lignocellulosic material (agroindustrial waste) assisted by the presence of microorganisms of the Clostridiaceae family. Currently, acetone—butanol—ethanol (ABE) fermentation, isopropanol-butanol-ethanol (IBE) fermentation, as well as glycerol fermentation present certain benefits since they can be derived from waste products. The growing interest in butanol fermentation is due to the fact that, in addition to being used as a chemical product, it can also be used as a biofuel.

5.3.1 Metabolic pathway of acetone-butanol-ethanol fermentation

The metabolic pathway involved in glucose fermentation includes different and interesting stages. Glucose is broken down by enzymes, commonly amylase, into fatty acids by anaerobic fermentation (Ezeji et al., 2010; Huang et al., 2010). In a very general way, it has been possible to visualize the mechanism of ABE fermentation. Starting from six or five carbons (pentose or hexose) a metabolism process is carried out via Embden-Meyerhof to produce pyruvate. A general balance indicates that from 1 moL of sugar, 2 moL of pyruvate and a net formation of two molecules of adenosine triphosphate (ATP) are produced.

Subsequently, pyruvate is converted into acetyl-CoA and CO_2. Eventually, the acetyl-CoA is converted to other intermediates,

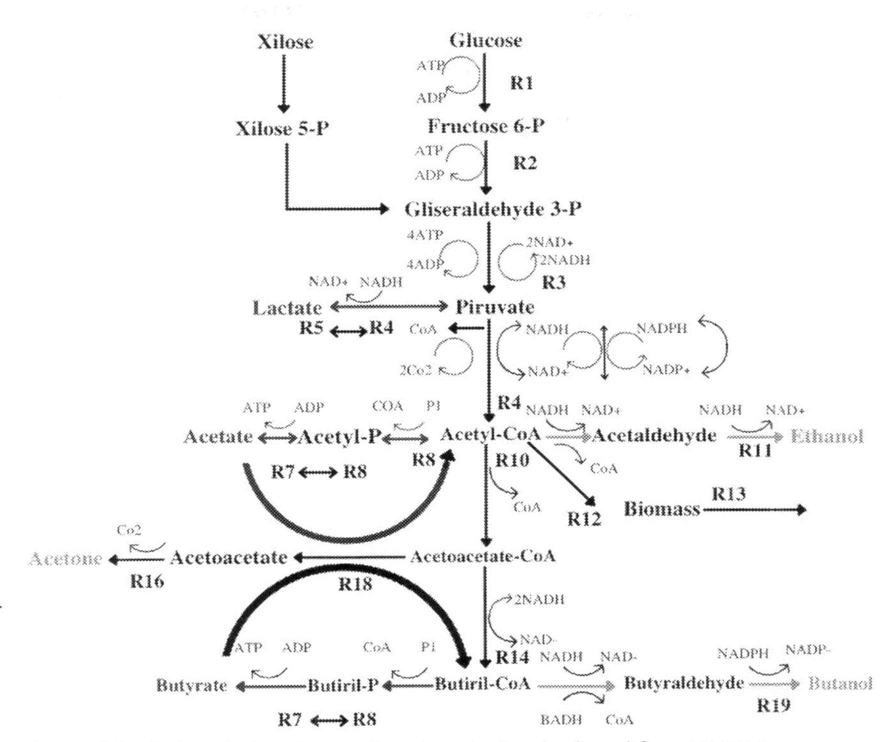

Figure 5.3 Biochemical pathway of acetone-butanol-ethanol fermentation.

acetaldehyde and butyraldehyde. Finally, these compounds are oxidized to generate the final products (acetone and acetate) and reduced to generate butanol and ethanol. The first stage is known as acidogenesis and occurs under specific pH conditions and iron restrictions. Consequently, ATP molecules are continuously produced during the process.

The second stage of ABE fermentation is solvent genesis. At this stage, the acids are re-assimilated to produce acetone, butanol, ethanol, acetic acid, butyric acid, hydrogen, and CO_2 as the main products. Fig. 5.3 schematically illustrates this.

5.3.2 Conventional raw material to produce butanol

Biobutanol can be generally produced from sugars. The origin of the fermented sugar determines and classifies the type of raw material used. Biobutanol produced from food crops is considered first-generation. Second-generation raw materials are considered to be those that do not compete directly with human consumption, that is, they are not

Table 5.3 Different generations of biofuel: source and process (Ndaba et al., 2015).

Generation	Feedstock	Processing technology	Examples of biofuel
First	Food crops, edible oil seed, animal fat	Esterification and transesterification of oils, fermentation of sugars, thermochemical process	Biodiesel, bioalcohols
Second	Waste cooking oil, nonedible oil seeds, lignocellulosic feedstock, agricultural waste	Physical, chemical, biological pretreatment of feedstock and fermentation, thermochemical process	Biodiesel, bioalcohols, syngas
Third	Algae	Algae cultivation, harvesting, oil extraction, transesterification or fermentation, thermochemical process	Biodiesel, bioalcohols, syngas, methane biohydrogen
Fourth	Algae and other microbes	Metabolic engineering of algae increasing carbon entrapment ability	Biodiesel, bioalcohols, syngas, methane biohydrogen

fundamentally food or do not compete for the use of land or cultivation areas. Any agro-industrial waste that has not been used in an initial harvesting stage can be classified as second-generation raw material. Recent research positions algae as third- and second-generation biomass, respectively, due to the low presence of lignin in the primary structure of algae. Table 5.3 shows some raw materials for various generations of fuels, as well as their process for biofuel production.

5.3.2.1 *First-generation biobutanol*

The production process to generate first-generation biobutanol is a relatively simple process, based on the fermentation of six-carbon sugars. The sugars are commonly obtained from a hydrolysis process of starch-rich crops such as maize, wheat, rice, and cassava. The hydrolysis process does not immediately generate hexose, initially dextrose is obtained, which is eventually converted to glucose through enzymes. The yields associated with this process are relatively high. For example, yields of 6.66 g L^{-1}, 0.18 g g^{-1} and productivities of 0.96 g Lh^{-1} have been reported using *C. beijerincki* (Li et al., 2014) producing primarily butanol. Slightly higher yields have been reported using a modified *C. beijerinckii BA101* strain, for example, 26 g L^{-1} to produce mainly acetone and butanol (Ezeji et al., 2004). On the other hand, other equally interesting strains have been used, for example, using *C-acetobutylicum DP 217* yields of 574.3 g L^{-1} have been recorded in an ABE fermentation (Li et al., 2014).

5.3.2.2 Second-generation biobutanol

As mentioned, the term "second-generation" is attributed to biofuels produced from biomass that is not for human consumption. The advantage of using biomass, usually agricultural residues, is the low cost, the abundance, as well as the net emissions of greenhouse gases.

Multiple attempts have been made to generate processes with high yields using various strains. For example, in an attempt to make the *C. acetobutylicum* strain directly usable cellulose, the gene from the *C. cellulovorans* strain was introduced into the *C. acetobutylicum* (Jin et al., 2011). Although several genes have been identified that can enhance the performance of the butanol production process, further research is still needed. The need to increase the overall yield of the process is mainly since second-generation biofuels require more sophisticated equipment, higher investment per unit of production and larger-scale facilities.

In recent years, several studies have aimed at producing butanol from second-generation biomass. Batch type processes have been reported, for example, the fermentation of liquid barley silage with butanol yields of 0.20 g g^{-1} of the monosaccharide used (Ndaba et al., 2015). On the other hand, Gottumukkala et al. (2013) used hydrolyzed rice straw for biobutanol production using the *Clostridium sporogenes BE101* strain. The strain generated butanol yields close to 3.43 g L^{-1}. Another interesting substrate was the spoilage date palm fruits fermented by *Clostridium acetobutylicum ATCC 824*. ABE fermentation reported a yield of 21.56 g L^{-1} (Komonkiat and Cheirsilp, 2013).

Considering biodiesel derived crude glycerol as raw material, Khanna et al. (2013) reported a yield of 5 g Lh^{-1} via anaerobic fermentation using *Clostridium pasteurianum* strains. In the same work it is mentioned the development of a mutant strain of *C. pasteurianum* in order to tolerate higher concentrations of butanol.

Although the use of biomass to generate second-generation biofuels has several advantages over the first-generation raw material, there are still obstacles to overcome. For example, in some geographical regions the cultivation spaces are insufficient to generate the necessary residues. On the other hand, processing time, as well as storage, remains a challenge. Third- and fourth-generation raw materials can be an alternative to overcome some of these limitations.

5.3.2.3 Third- and fourth-generation biobutanol

Algae has been recently projected as a raw material of high interest. Few reports have been made where butanol is produced from algae. The primary reason for this behavior is that the main content of algae (approximately 50%) is oil, making them perfect candidates for biodiesel production. Once all the oil has been extracted, the waste generated could be used to produce biobutanol. Currently, there are two types of algae, microalgae and macroalgae. Microalgae are made particularly from microscopic unicellular organisms; on the other hand, macroalgae contain multiple cells with a structure similar to that of rooted plants. Macroalgae are categorized into red, green, and brown, depending on the color. The specific characteristics of seaweeds are low protein and lipid content, but high carbohydrate content (Castro et al., 2015; Cheng et al., 2015; Van der Wal et al., 2013).

Regarding the production process, Castro et al. (2015) reported the use of algae in a hydrolysis process to release sugars and its later use for ABE fermentation using *C. Saccharoper butylacetonicum N1−4*. The methodology reported concentrations of 3.74 g L^{-1} of butanol (Ellis et al., 2012). The study by Castro et al. (2015) demonstrated the use of almost 100% biomass for biofuel production, however, it also showed the need to improve the productivity of the process.

In a similar process, Van der Wal et al. (2013) presented a study on the use of algae for the production of acetone, butanol, and ethanol. Using *C. beijerinckii*, a productivity of 0.35 g g^{-1} was reported.

5.3.2.4 Problems associated with acetone−butanol−ethanol fermentation

Even though there is an extensive record of useful raw material as well as microorganisms that can successfully ferment into butanol, there are several obstacles to ABE fermentation. These obstacles must be overcome in order to compete directly with fossil fuels. These barriers can be divided into three categories (Jones and Woods, 1986):

1. Low concentration of butanol (less than 20 g L^{-1}) due to the inhibition of microorganisms. At higher concentrations, the microorganisms present a toxic reaction to butanol.
2. Low performance due to heteroformation. Ranges between 0.28 and 0.33 g g^{-1}.
3. High cost associated with butanol recovery.

5.3.3 Isopropanol-butanol-ethanol fermentation

In the 1980s, several generic engineering tools were used to improve the productivity of *Clostridium* strains, which triggered IBE fermentation. Most studies were conducted at Delft University of Technology in the Netherlands.

While ABE fermentation is interesting because of the formation of several useful components, there is a problem associated with acetone production. For example, if the biobutanol produced reached the same amount of bioethanol generated in Brazil, there would be twice as much acetone as required globally (dos Santos Vieira et al., 2019). Additionally, acetone does not qualify for the incentives that some governments apply to local industries. Therefore, likely unfavorable prices are expected in case some ABE plants are installed to meet certain demand. As mentioned, some alternatives depend on metabolic engineering, for example, (1) decrease or eliminate acetone production, seeking higher yields for butanol in ABE fermentation (Zheng et al., 2015) and (2) develop some acetone-free strain in butanol production. Acetone production can be avoided by using *C. pasteurianum* strains that can convert glycerol to 1,3-propanediol. However, in a scenario of high demand for butanol, an oversupply of 1,3-propanediol would be generated, with a demand of around 0.25 million tons in 2020 (Biddy et al., 2016).

Other ways to reduce acetone production are through its conversion to isopropanol. One biological route to achieve this is through strains that can act on the solventgenic stage to reduce acetone to isopropanol in the so-called IBE fermentation. This alternative has the advantage that all compounds derived from the fermentation can be used directly as fuel. However, the production of IBE compounds is more sensitive to inhibition by products and consequently less efficient than ABE fermentation (Bankar et al., 2013; Zhang et al., 2018).

In that sense, the development of new *Clostridium* strains has been motivated by the poor performance of wild-type strains. For example,

Table 5.4 Performance of isopropanol-butanol-ethanol batch fermentation according to carbon source and *Clostridium* strain (dos Santos Vieira et al., 2019).

Carbon source	g L^{-1}	Strain	Butanol (g L^{-1})	Isopropanol (g L^{-1})
Glucose	20	*C. beijerinckii VPI 2968*	3.3	0.6
	20	*C. aurantibutyricum NCIMB 10659*	3.1	0.6
	20	*C. beijerinckii VPI 2432*	6	1.6
	60	*C. beijerinckii BGS1*	10.2	3.4
Glucose and xylose	40/20	*C. beijerinckii NRRL B593*	6.9	3.2
Manose	20	*C.* sp. *A1424*	4.4	1.9
Fructose	20	*C.* sp. *A1425*	4.4	1.6
Cellobiose	20	*C.* sp. *A1426*	4.5	2.6
Sucrose	20	*C.* sp. *A1427*	5	1.7
	60	*C.* sp. *A1428*	9.8	2.5
Caen molasses	30	*C. beijerinckii optinoii*	7.6	4.6
Cassava bagasse hydrolysate	30	*C. beijerinckii ATCC 6014*	8.2	4.3
Coffe silverskin hydrolysate	21	*C. beijerinckii DSM 6423*	4.4	2.2
Birchwood xylan	60	*C. beijerinckii NJP7*	2.1	0.5

in a batch culture, using *Clostridium* bijerinckii, it is difficult to exceed 6 g L^{-1}. A recent study indicates that the best result obtained so far was using *C. beijerinckii* BSG1, with a production of 10.2 g L^{-1} of butanol and 3.4 g L^{-1} of isopropanol (Zhang et al., 2018). Table 5.4 shows several strains of the *Clostridium* family that have been tested for IBE fermentation.

Finally, concerning the raw material that can be used in IBE fermentation, due to the fermentative capacity of five- and six-carbon sugars, the raw materials can be considered relatively similar.

5.4 Process intensification applied to butanol production

Process intensification is a philosophy that has been subject to numerous discussions and interpretations. Similarly, various definitions have been proposed in the literature, depending on how the researcher perceives the intensification of processes.

For example, Arizmendi-Sánchez and Sharratt (2008) consider the intensification of processes as the synergistic integration of processes and phenomena. On the other hand, Huang et al. (2007) define PI as the development of innovative devices and techniques that offer drastic improvement in chemical manufacture and processing. Additionally, they mention that it should substantially decrease the volume of the equipment, the energy consumption, the waste formation, leading to a cheaper, safer, and sustainable process technologies.

However, it is evident that through the use of these types of philosophies, the improvement of the processes is evident. In the following paragraphs, we will describe how process intensification has greatly benefited the production of butanol. The information will be divided into two sections, process intensification applied to the reaction stage, and then process intensification applied to the purification section.

5.4.1 Process intensification in the reactive zone

The improvement in biobutanol production can be achieved through two strategies (1) a biological approach, that is, trying to produce strains for a hyperbutanol production (Jang et al., 2012), and (2) the theoretical optimization and modeling of more efficient hybrid processes.

As mentioned, the main reason to improve this process is the inhibition effect caused by the concentration of butanol, since with concentrations close to 15 g L^{-1} (Quiroz-Ramírez et al., 2018). In order to improve the process globally, several options have been proposed. The vast majority of options focused on the reaction section propose the integration of the reaction with an in-situ recovery process.

The objective of in-situ recovery techniques is to remove the products as soon as they are formed, this increases the productivity, as well as the overall concentrations of the fermentation by avoiding inhibition processes.

To compare various in-situ techniques, the amount of substrate used, productivity, yield, and so on are used. In the following paragraphs some techniques used will be presented, as well as the general characteristics presented in various research works.

5.4.1.1 Gas stripping

Gas stripping is a separation technique that involves the removal of various compounds by dissolution in a gas that passes through the fermentation broth. In the particular case of ABE fermentation, it is necessary to recycle the fermentation gases (CO_2 and H_2) or to apply anaerobic gases, for

Table 5.5 Acetone–butanol–ethanol fermentation performance with recovery by gas stripping (Outram et al., 2017).

Mode	Strain	ABE productivity g (ABE Lh^{-1})
Batch	*C. acetobutylicum P262*	0.31
	C. acetobutylicum P262	0.32
	C. beijerinckii BA 101	0.6
	C. beijerinckii BA 101	0.31
	C. beijerinckii BA 101	0.4
	C. beijerinckii CC101	0.17
	C. beijerinckii NRRL B593	0.29
Fed-batch	*C. beijerinckii BA 101*	1.16
	C. beijerinckii BA 101	0.59
	C. acetobutylicum P262	0.26
Continuos	*C. beijerinckii BA 101*	0.92
	C. beijerinckii NRRL B593	0.93
	C. beijerinckii NRRL B593	1.3

example, nitrogen-free oxygen (Ennis et al., 1987, 1986) in the fermenter to remove ABE compounds from the gas stream (Qureshi et al., 1992). The advantage of gas stripping is that it does not require expensive equipment or substantial modification to the plant, that is, gas stripping is considered a relatively simple technique.

A large number of studies have been conducted evaluating gas stripping, especially from Professor Ezeji's working group (2007, 2004, 2003). In general, these studies conclude that the use of gas stripping significantly increases fermentation productivity. Table 5.5 presents some works where gas stripping is used

There are some results in Table 5.5 that show higher yields than those theoretically reported in the literature. Ezeji et al. (2013, 2003) mention that the increase may be due to the consumption of other carbon sources present in the culture medium, for example, sodium acetate. In some works, a decrease in yield has been reported compared to a nonintegrated gas stripping operation. For example, Ennis et al. (1986) reported a 31% decrease in yield. This decrease was attributed to inefficient condensate capacity. That is, not all components are captured and consequently not quantified for the calculation of the yield.

The gas stripping technique has some limitations due to the low concentration of ABE compounds, the large amount of water removed and the large amounts of gas flow required (Xue et al., 2012).

5.4.1.2 Vacuum fermentation

Vaccum fermentation is the reduction of pressure in the fermenter, causing the ABE compounds to "boil off" at the temperature of fermentation. This technique was initially used in the ethanol industry to selectively remove ethanol from the fermentation broth.

Mariano et al. (2012a, b) demonstrated the viability of this technique applied to ABE fermentation in laboratory scale. The system was initially modeled, generating concentrations in the range of 5–15 g butanol L^{-1}. However, the experimental result was lower than expected.

There are two different operation modes for vacuum fermentation, constant and cyclic. The cyclic vacuum fermentations are considerably more competitive in terms of energy demand than conventional distillation. The cyclic vacuum process allows the concentration of butanol to build up, then reduces the concentration rapidly by applying a vacuum for 2 h, repeating this process throughout the fermentation (Mariano et al., 2012a, b).

In an investigation Mariano et al. (2012a, b), the energy requirement was evaluated by adding vacuum to the fermentation. Their research group observed an energy reduction of about 11.2 MJ kg^{-1} of butanol for a continuous vacuum and 11 MJ kg^{-1} for an intermittent vacuum.

In general, the application of vacuum can increase substrate use as well as fermentation productivity. However, without an efficient capture process, it is difficult to observe.

5.4.1.3 Pervaporation

Pervaporation uses a membrane between the fermentation broth and the gas phase, a simplified scheme is shown in Fig. 5.4.

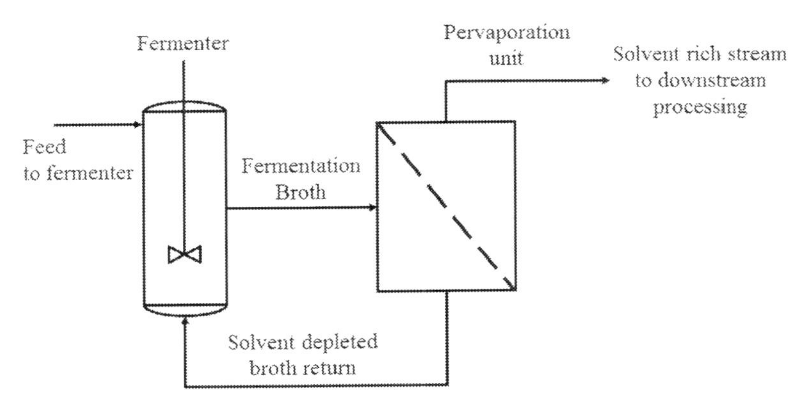

Figure 5.4 Diagram of in-situ fermentation combined with pervaporation.

The pervaporation process generates a more complex scheme, so an external unit is required. Pervaporation may be integrated with the bioreactor (Larrayoz and Puigjaner, 1987), however, this is a very unusual configuration (Outram et al., 2017).

5.4.1.4 Liquid–liquid extraction

Liquid–liquid extraction (LLE) is a technique widely used in the industry. LLE focuses on exploring the relative difference in solubilities of a component in two immiscible components. In general, the solvent used is an organic liquid immiscible in water, so when used in a fermentation process, the product will be transferred preferably from the aqueous phase to the organic phase. One of the greatest difficulties in LLE is to find the extractant that has the necessary characteristics of the extractant: nontoxic for the microorganism, with a high partition coefficient, immiscible with water and that does not form emulsions, low viscosity, chemically stable at high temperatures, sterilizable, commercially available.

Ishii et al. (1985) and Roffler et al. (1987) conducted several tests to find a suitable extractant, and both concluded that oleic alcohol is an acceptable extractant for butanol extraction. Some other solvents have been tested, for example, decanol, dibutylphthalate, 2-butyl-1-octanol, and poly(propylene glycol) 1200 (Bankar et al., 2012; Barton and Daugulis, 1992; Eckert and Schügerl, 1987; Evans and Wang, 1988; Qureshi and Maddox, 1995; Wayman and Parekh, 1987). However, decanol is toxic to bacteria.

By using LLE, the yields obtained are low, compared to gas stripping. The average yield using LLE is 0.25 gABE g^{-1} of substrate, which is approximately 40% lower than the evaporation techniques. In order to generate an economically viable process, it is necessary that the extractant be recyclable, thus requiring the removal of ABE to be direct. Table 5.6 shows various extractants used with various microorganisms.

5.4.1.5 Adsorption

Adsorption is a surface phenomenon between an adsorbent and an adsorbate. Adsorption is the oldest technique for recovering ABE in situ. A wide variety of adsorbents has been tested for the recovery of ABE. Among the most commonly used are activated carbon (Groot and Luyben, 1986; Xue et al., 2016), zeolites or silicates(Ennis et al., 1986; Remi et al., 2012; Xue et al., 2016) and polymer resins (Nielsen and Prather, 2009; Nielsen et al., 1988). In initial conclusions, Qureshi et al.

Table 5.6 Acetone–butanol–ethanol fermentation with in-situ recovery by LLE (Outram et al., 2017).

Mode operation	Microorganism	Substrate (g L^{-1})	Yiled (g ABE/g substrate)	Extractant
Bacth	*C. acetobutylicum ATCCC824*	Glucose (97)	0.17	kerosene
	C. acetobutylicum ATCCC824	Glucose (78)	0.17	50%wt dodecanol in kerosene
	C. acetobutylicum ATCCC824	Glucose (89)	0.16	30%wt tetradecanol in kerosene
	C. acetobutylicum ATCCC824	Glucose (100)	0.19	oley alcohol
	C. acetobutylicum ATCCC824	Glucose (100)	0.17	50%wt oley alcohol in decane
	C. acetobutylicum ATCCC824	Glucose (100)	0.18	50%wt oley alcohol in benzyl benzoate
	C. Saccharoper butylacetonicum N1–4	Potato glucose (75)	0.38	oley alcohol
	C. Saccharoper butylacetonicum N1–4	Potato glucose (74)	0.4	methylated crude palm oil
	C. Acetobutylicum BCRC10639 (ATCC824)	Glucose (unknown)	0.21	biodiesel
Fed Batch	*C. Acetobutylicum (ATCC824)*	Glucose (86)	0.36	oley alcohol
	C. Acetobutylicum (ATCC824)	Glucose (117)	0.37	oley alcohol
	C. Acetobutylicum (ATCC824)	Glucose (155)	0.24	oley alcohol
	C. Acetobutylicum (ATCC824)	Glucose (218)	0.22	oley alcohol
	C. Acetobutylicum (ATCC824)	Glucose (303)	0.21	oley alcohol
	C. Acetobutylicum (ATCC824)	Glucose (86)	0.23	PPG1200
	C. Acetobutylicum BCRC10639 (ATCC824)	Glucose (unknown)	0.31	Biodiesel

(2005) highlights that the use of adsorbents improves the fermentation process from 5 g butanol to 810 g butanol L^{-1}; however, recently a trend toward polymer resins has been observed.

Most adsorbents have not been tested in conjunction with ABE fermentation; instead, several models have been generated. Yang et al. (1994) are one of the few research groups that have worked with adsorbents in conjunction with fementation. They demonstrated that the use of 30% resin at fermentation can achieve a 130% increase in fermentation productivity. When this technology was adapted to fed-batch fermentation with an external column, productivity increased by 233% in a single adsorption cycle.

Lee et al. (2015) added adsorbent directly to the fermenter without observing fouling. This work demonstrated that in batch fermentation, using *C. acetobutylicum* ATCC 824, the concentration of product can be increased due to product adsorption, reducing toxicity. The final concentration of butanol was 10 g butanol L^{-1}. Wiehn et al. (2014) proposed an expanded adsorption bed process. This allowed the fermentation broth to pass through the adsorbent without being separated from the microorganisms, due to the empty space in the bed. A reduction in biomass concentration was observed compared to the control, however, an increase in yield and productivity of 14%–65% respectively was observed.

Weizmann et al. (1948) observed that butanol is preferentially adsorbed compared to acetone, followed by ethanol. Table 5.7 shows some in-situ adsorption processes applied for the capture of ABE.

A brief comparison of ABE's intensified onsite recovery techniques is possible. Considering batch fermentation, all techniques generated a

Table 5.7 Acetone–butanol–ethanol fermentation with in-situ recovery by adsorption (Xue et al., 2016; Yang et al., 1994).

Mode operation	Microorganism	Substrate (g L^{-1})	Yield (g ABE/g substrate)
Bacth	*C. acetobutylicum ATCCC824*	Glucose (92)	0.32
Fed batch	*C. acetobutylicum ATCCC824*	Glucose (190)	0.32
	C. acetobutylicum ATCCC824	Glucose (199)	0.32
	C. acetobutylicum ATCCC824	Glucose (180)	0.28
	C. acetobutylicum JB200	Glucose (158)	0.22

positive effect on fermentation; this is due to the continuous removal of butanol that decreases inhibition, allowing better use of substrate and the possibility of prolonging fermentation. However, according to what has been previously reported, the most promising technique for batch fermentation is gas stripping. Once the inhibition effect can be overcome using some technique, the next step is to increase the substrate load and the fermentation time to obtain better productivities. One of the key factors for applying enhanced onsite remission techniques in ABE fermentation is the potential for reducing energy consumption due to the increased concentration of the flow that will be sent to the purification process.

5.4.2 Process intensification in the downstream process

Because of the low concentration obtained from fermentation, product purification still poses a huge challenge associated with the development of biofuels. An acetone/butanol/ethanol ratio of 3:6:1 is presented in a classical mixture from fermentation; on the other hand, the amount of water depends on many factors associated with the upstream phase. In addition to the difficulty of separating such low concentrations of biobutanol, the interaction of the components in an ABE mixture makes the process of separation more complicated. The ternary diagram in Fig. 5.5 depicts two azeotropes, a homogeneous azeotrope between ethanol and water, and a heterogeneous azeotrope between biobutanol and water, to demonstrate this.

Several recovery techniques have been suggested nowadays, we might highlight distillation, adsorption, LLE, membrane distillation, and gas stripping from those techniques with further study. The maturity of distillation and its popularity for mixture separation is not hard to understand. With this in mind, multiple research groups have suggested it as a potential solution to purify the ABE mixture. In their idea, a bioreactor creates a vacuum chamber with a flowing broth. As a consequence, this process allows 30–37 $g \cdot L^{-1}$ to be obtained. In order to recover biobutanol from a butanol-water mixture, Luyben (2012) has suggested a pressure-swing azeotropic distillation method. To separate the liquid mixture, two columns with a different range of pressures are used in this work. For biobutanol recovery, Matsumura et al. (1988) used distillation columns. The energy requirements of 79.5 MJ kg^{-1} to separate the mixture were stated by this research group. A collection of four alternatives to purifying all the components in the ABE mixture has been stated by Van der Merwe et al.

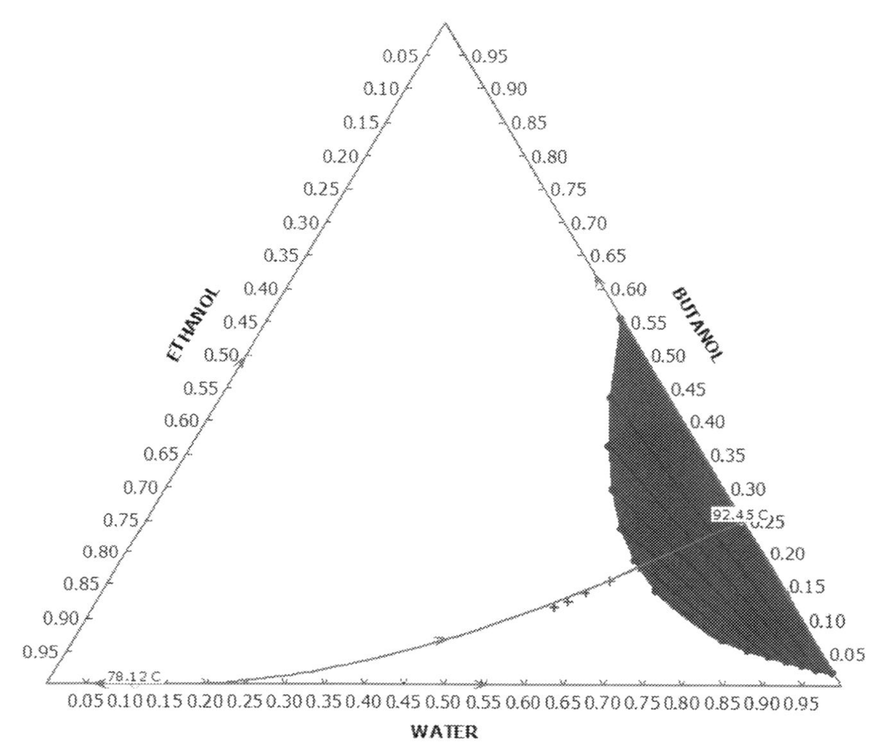

Figure 5.5 Ternary diagram for the acetone–butanol–ethanol mixture (mole basis).

(2013). Most schemes are based on columns of distillation. A flash fermentation integrated with a continuous process for ABE recovery was proposed by Mariano et al. (2011). In their proposal, a bioreactor creates a vacuum chamber with a circulating broth. However, those studies have concluded that conventional purification options do not represent a real solution for reducing the energy requirements associated with the purification of the ABE mixture. Table 5.8 shows the energy requirements when using conventional separation processes.

The application of process intensification to the various parts of the production process is an area of opportunity. Process intensification typically means strategies for reducing the size of process equipment, the cost of capital and waste generation, improving energy quality and inherent protection (Qian et al., 2018).

Grisales-Díaz and Olivar-Tost (2017) proposed four processes of heat-integrated distillation were proposed to reduce the energy requirements of

Table 5.8 Energy requirements of several purification technologies (Groot et al., 1992; Hugo et al., 2016).

Recovery system	Energy requirement ($MJ\ kg^{-1}$-product)
Distillation	12.8
Distillation	16.7
Distillation	15.2
Steam distillation	21
Gas stripping	18.9
Adsorption	7.1
Pervaporation	11.9
LLE	7.7
Vacuum evaporation	21.8
Double-effect distillation	8
Pervaporation	9.6
Vacuum evaporation	10.8
Adsorption	36.7
Gas stripping	23.3
LLE	15.6
Pervaporation	10
Distillation	5.2
Double-effect distillation	3.4
Double-effect distillation	5.7

ABE recovery. Using the fermentation broths of several biocatalysts, the energy requirements and economic evaluations were carried out. Processes with four distillation columns and three distillation columns (between 7.7 and 11.7 MJ fuel kg^{-1}-ABE) had similar energy requirements. The most economical (0.12–0.16 \$ kg^{-1}-ABE) method was the double-effect system (DES) with four columns. DES achieved energy requirements ranging from 6.1 to 8.7 MJ fuel kg^{-1}-ABE for ABE recovery from dilute solutions. The lowest energy consumption (between 4.7 and 7.3 MJ fuel kg^{-1}-ABE) was achieved by vapor-compression distillation (VCD). ABE recovery DES and VCD energy requirements were lower than integrated reactors. Energy requirements for the production of ABE were 1.3–2.0 times greater than for alternative biofuels (ethanol or isobutanol). However, due to the production of hydrogen during the fermentation of ABE, the energy efficiency of the production of ABE was equal to that of ethanol and isobutanol (between 0.71 and 0.76).

From the same research group, but considering IBE fermentation (Grisales-Díaz and Olivar-Tost, 2017), A combination of azeotropic and extractive distillation was the new distillation method suggested in their work for IBE dehydration. Butanol has broken the azeotropic activity of isopropanol-water and ethanol-water in a complex reflux system, without an additional entrainer. Butanol-water azeotrope refused decantation. Using Aspen Plus tools, the distillation device was assessed. An energy requirement of between 6.5 and 8.2 MJ-fuel kg^{-1}-IBE has been achieved by the alternative distillation method. The fuel requirement was reduced to 3.4–4.1 MJ-fuel kg^{-1}-IBE using steam compression distillation. The IBE dehydration energy efficiency was between 0.72 and 0.79. The recovery of IBE has been linked to alternative biofuels. The IBE energy demand was 0.92–1.4 and 1.4–2.4 times greater than the dehydration of isobutanol and ethanol, respectively. However, through the development of hydrogen, IBE fermentation achieved an equal energy efficiency.

A hybrid intensified alternative, combining a conventional distillation with a membrane, was proposed by van Wyk et al. (2018). First, using a polydimethylsiloxane membrane and a model solution of ABE as the feed, an experimental investigation was performed. Different operating temperatures were examined and the most promising results were seen at 40°C. Second, using Aspen Plus, the experimental results were used as an input for process simulations. Two separation schemes for ABE, one consisting only of distillation (conventional process) and one with an upstream pervaporation unit followed by an alternative distillation scheme, were studied. The butanol concentration after pervaporation was sufficiently high for the proposed pervaporative method so that it could be further concentrated through a liquid–liquid separation at the beginning of the separation train. The results of the simulations showed that the traditional system was the most energy-consuming, and that energy consumption was reduced by 53% by integration with an upstream pervaporation unit. For the distillation system, the energy requirement was 33.3 MJ kg^{-1} butanol, whereas the pervaporation-distillation scheme was 15.7 MJ kg^{-1} butanol.

In this work, to reduce the energy requirements of ABE recovery, four processes of heat-integrated distillation were proposed. The energy requirements and economic evaluations were performed using the fermentation broths of several biocatalysts. Energy requirements of the processes with four distillation columns and three distillation columns were similar (between 7.7 and 11.7 MJ fuel kg^{-1}-ABE). DES with four columns was the most economical process (0.12–0.16 \$ kg^{-1}-ABE). ABE recovery from dilute solutions by DES achieved energy requirements between 6.1 and 8.7 MJ fuel kg^{-1}-ABE. VCD reached the lowest energy consumptions (between 4.7 and 7.3 MJ fuel kg^{-1}-ABE). Energy requirements for ABE recovery DES and VCD were lower than that for integrated reactors. The energy requirements of ABE production were between 1.3- and 2.0-fold higher than that for alternative biofuels (ethanol or isobutanol). However, the energy efficiency of ABE production was equivalent to that for ethanol and isobutanol (between 0.71 and 0.76) because of hydrogen production in ABE fermentation.

Sánchez-Ramírez et al. (2017b) propose the use of LLE based hybrid separation combined with dividing-wall column (DWC) technology for the purification of the mixture of ABE. The proposed configurations are the product of multiobjective optimization aimed at identifying designs that satisfy the tradeoff among three goals: the total annual cost and the eco-indicator 99, as economic and environmental objectives; and in order to forecast the dynamic properties, it was included the condition number as controllability objective. Those three objectives were simultaneously optimized using a hybrid stochastic optimization algorithm. Among the four designs, the best economic output and reasonably good values of condition number and eco-indicator 99 were demonstrated in the scheme where only a reboiler is included.

A new method for butanol recovery by improved distillation (e.g., DWC technology) is proposed by Patraşcu et al. (2017) using only a few operating units in an optimized series to minimize overall costs. A 40 ktpy butanol plant capacity is considered and 99.4 %wt butanol, 99.4 %wt acetone, and 91.4 %wt ethanol purity are considered.

Further, from the same research group, Patraşcu et al. (2018) presented an intensified alternative to overcome the high downstream processing costs. Their study proposes a novel intensified separation process based on a heat pump (vapor recompression)-assisted azeotropic dividing-wall column (A-DWC). Pinch analysis and rigorous process simulations have been used for the process synthesis, design, and optimization of this novel sustainable process. Remarkably, the energy requirement for butanol separation using heat integration and vapor recompression-assisted A-DWC is reduced by 58% from 6.3 to 2.7 MJ kg^{-1} butanol.

Finally, another example of intensified hybrid process is shown next. Combining traditional distillation schemes with LLE is a reasonably well-studied option. An interesting alternative seems to be LLE, as an organic water-immiscible extractant agent may be used to prevent all azeotropes from developing. In order to purify an ABE mixture at high biobutanol

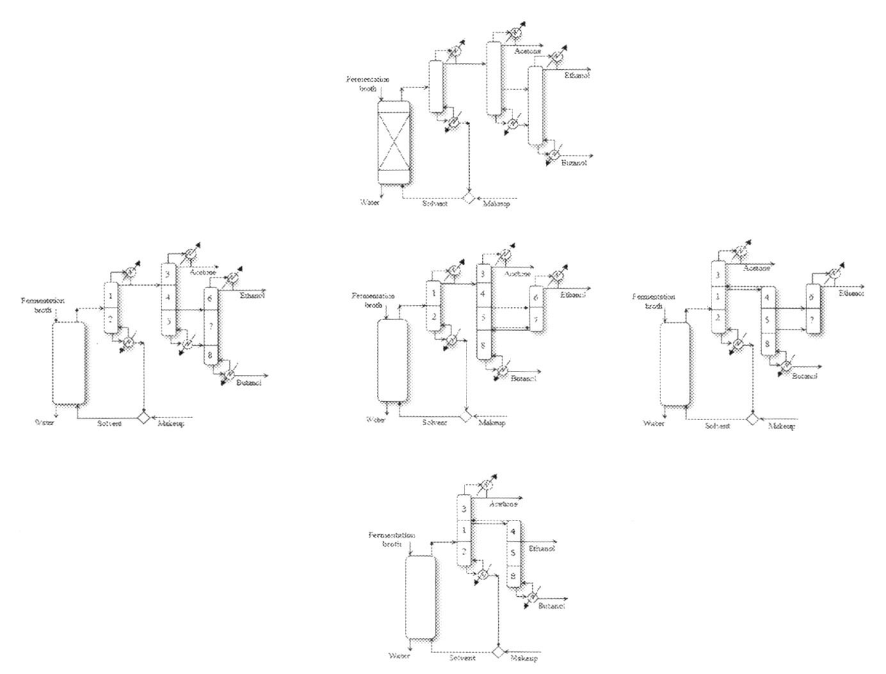

Figure 5.6 Intensified alternatives of biobutanol purification.

concentration, Errico et al. (2016) suggested several hybrid separation processes; the hybrid configuration consists of a liquid–liquid extractive column, followed by thermally coupled distillation columns, producing many alternatives with relatively low energy consumption. Fig. 5.6 shows the separation schemes developed from a hybrid case base. In general terms, the technique for forecasting new alternative schemes is based on the implementation of thermal couplings, column section transposition and intensification of processes. Later, Errico et al. (2017) also suggested a range of hybrid alternatives based on extractive columns of liquid–liquid coupled with columns of dividing walls. Fig. 5.7 shows some examples of intensified scheme considering DWC alternatives.

Both works were developed under an optimization framework, having as objective function the total annual cost, the eco-indicator 99 and the condition number as economic, environmental and controllability indexes, respectively.

The findings reported in Table 5.9 indicate that all liquid-assisted DWCs have a lower value of total annual cost (TAC) from an economic

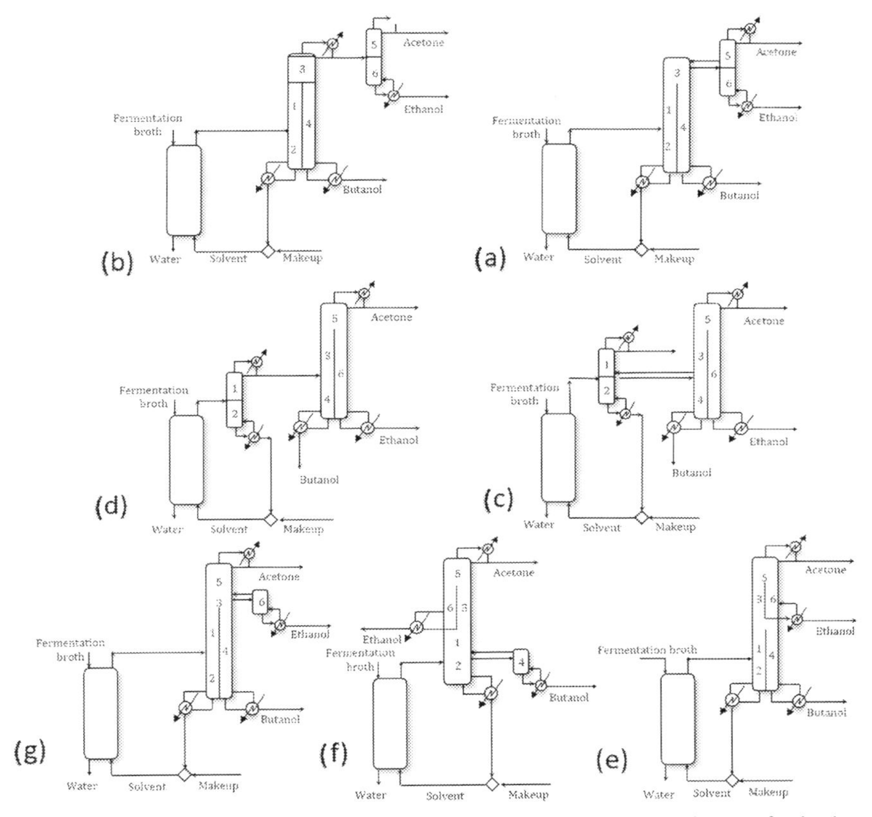

Figure 5.7 Intensified alternatives based on dividing-wall column schemes for biobutanol purification.

point of view in regard to the best simple column distillation reported Errico et al. (2017). The configuration of Fig. 5.6a showed the lowest values of all the objective functions among the liquid-liquid-assisted traditional DWCs. It realized a 13.6% decrease in TAC and better controllability, but a penalty of 12.5% is noted in EI99 in comparison with a conventional process previously reported. The penalty is because of the highest consumption of DWC utilities.

The TAC and the EI99 have been penalized as compared with a conventional case when the solvent was retrieved as the lowest stream in the DWC. Regarding the output data of Fig. 5.7, the economic output of Fig. 5.7g is predicted because the solvent and ABE mix were segregated in the same column as the more intensified alternative. The configuration of Fig. 5.7e

Table 5.9 Objective function of the evaluated schemes in Fig. 5.7.

Objective function	Figure (a)	Figure (b)	Figure (c)	Figure (d)	Figure (e)	Figure (f)	Figure (g)
TAC [k$ yr^{-1}]	108.54	105.57	115.5	101.78	100.85	100.59	97.88
EI99 [kpoints yr^{-1}]	13.73	12.93	14.34	13.3	12.79	14.74	12.22
CN	1402	1.7	$1.22\eta10^{17}$	3.9	7.3	9888.3	18,994.4

Table 5.10 Design parameters of the scheme.

	Extractor	1 + 2	4	5 + 3	6
Number of stages	5	43	43	71	7
Feed location	—	13	—	—	—
Reflux ratio	—		—	0.644	—
Distillate flowrate [kg h^{-1}]	—	—	—	7.717	—
Residue flowrate [kg h^{-1}]	—	712.108	13.681	—	0.316
Liquid split flowrate [kg h^{-1}]	—	43.463	17.383	—	—
Vapor split flowrate [kg h^{-1}]	—	—	—	—	—
Extract flowrate [kg h^{-1}]	733.873	—	—	—	—
Solvent flowrate [kg h^{-1}]	712.147	—	—	—	—
Diameter [m]	0.335	0.288	0.288	0.3788	0.299
Pressure [kPa]	101.3	101.3	101.3	101.3	101.3
Condenser duty [kW]	—	0	0	11.233	0
Reboiler duty [kW]	—	69.92	0.633	0	0.023

extends the relation to the best fluid-liquid extraction-assisted simple column distillation and enhances the efficiency of all three objectives. The TAC and EI99 were reduced in particular by 22%, 18%, and the better controllability index, respectively. Table 5.10 provides details about the setup of Fig. 5.7e.

5.5 Controllability studies applied to intensified alternatives for biobutanol purification

Although the work proposed by Sánchez-Ramírez et al. (2017b) showed some indications regarding the dynamics and controllability of some intensified schemes for biobutanol purification, the full dynamics of this type of process were not studied. That is, the study was made considering a strategy of controllability at zero frequency, not in the whole range of the frequency.

Few studies have addressed the full dynamics of this type of scheme. An example of this type of studies is the one made by Patraşcu et al. (2019); For ABE, they proposed an effective downstream process, enabling approximately 60% energy savings compared to a traditional separation series. In an A-DWC coupled with a heat pump and applying heat integration, this was feasible. The energy consumption was reduced to just 2.7 MJ kg^{-1} of butanol (or 7.5% of butanol's energy content), but the process must also be controllable in order to take maximum advantage of the energy savings provided by this highly integrated process. The dynamics and control of this novel A-DWC method used for the purification of biobutanol were discussed in their work.

An efficient downstream process has been developed for the ABE, allowing about 60% energy savings as compared to a conventional separation sequence. This is achievable in an A-DWC coupled with a heat pump and applying heat integration. The energy use is reduced to only 2.7 MJ kg^{-1} butanol (or 7.5% of the energy content of butanol), but to take full advantage of the energy savings offered by this highly integrated process, the process must be also controllable. This work presents the dynamics and control of this novel A-DWC process used for biobutanol purification. With Aspen Plus Dynamics, a rigorous, pressure-driven dynamic simulation has been established. The device can manage small and brief disturbances (e.g., 5% change in feed flowrate for 10 min) after adding the necessary PID controls (e.g., flow, pressure, liquid level, and temperature) to the base-case process, but if the disturbance continues, the process shuts down for a long time. The system's controllability was enhanced by installing an additional reboiler and condenser to regulate the temperature on the feed-side stripping section to solve this issue. In addition, adding concentration controls ensures that when feed flowrate or composition disturbances occur, the product purity is kept constant.

On the other hand, Sánchez-Ramírez et al. (2017a) studied the controllability on intensified alternatives for biobutanol purification. These alternatives provide energy and economic savings, in addition to reducing the effect on the environment. However, few studies have studied the process's control properties, which include the separation of a mixture of ABE. On several hybrid intensified distillation processes, including traditional, thermally coupled, thermodynamically equivalent and intensified designs, a controllability analysis using the singular value decomposition technique and a closed-loop dynamic analysis was conducted. The findings suggested that an intensified system that is incorporated with only two distillation columns rather than three distillation columns showed strong dynamic properties under the closed-loop control policy. Moreover, under open-loop analysis, thermally coupled sequence showed better control properties.

5.6 Conclusions

In this chapter, a complete analysis has been made on butanol, to a greater extent, to butanol from fermentative processes. The various stages involved in the overall production of butanol by fermentation have been analyzed separately, with great emphasis on the intensified technologies applied in the reactive process and in the purification process. During this in-depth analysis, some opportunities for process improvement were observed. While the conversion of biomass into biobutanol is technically feasible, there are still several areas of opportunity to make it commercially feasible to generate it on an industrial scale, making biobutanol a competitive fuel. First, it is important to exploit all potential sources of biobutanol to reduce the costs associated with raw materials. It is important to encourage the use of renewable materials as agricultural residues, since they give additional use to the residues, with the expectation of paying lower quantities for those raw materials. The low concentration and yield, along with the inhibition of microorganisms due to butanol, is another obstacle that must be overcome in the development of biobutanol. In the fermentation stage, therefore, changes are needed, for which genetic engineering plays a key role in improving the development of butanol. Alternative processing schemes have been proposed to increase yield and efficiency, in addition to the growth of modified microorganisms. Following downstream processes, due to the presence of homogeneous and heterogeneous azeotropes, the separation of the components in the ABE mixture from water is not a simple task. In addition, in downstream processes, due to the highly diluted products obtained during fermentation, there are high energy requirements. The upgrade of the separation train is therefore mandatory in order to reduce the energy needed to carry out the separation, thus preventing the use of extremely costly technologies. Once the various actors can overcome the limitations of each process stage, the butane produced by fermentative processes will be able to compete fairly, even surpassing, the synthetic process for obtaining butanol.

References

Arizmendi-Sánchez, J.A., Sharratt, P.N., 2008. Phenomena-based modularisation of chemical process models to approach intensive options. Chem. Eng. J. 135, 83–94.

Bankar, S.B., Survase, S.A., Singhal, R.S., Granström, T., 2012. Continuous two stage acetone-butanol-ethanol fermentation with integrated solvent removal using *Clostridium acetobutylicum* B 5313. Bioresour. Technol. 106, 110–116.

Bankar, S.B., Survase, S.A., Ojamo, H., Granström, T., 2013. Biobutanol: The outlook of an academic and industrialist. RSC Adv. 3, 24734–24757.
Barton, W.E., Daugulis, A.J., 1992. Evaluation of solvents for extractive butanol fermentation with *Clostridium acetobutylicum* and the use of poly(propylene glycol) 1200. Appl. Microbiol. Biotechnol. 36, 632–639.
Biddy, M.J., Scarlata, C.J., Kinchin, C.M., 2016. Chemicals from biomass: a market assessment of bioproducts with near-term potential. NREL Rep.
Castro, Y.A., Ellis, J.T., Miller, C.D., Sims, R.C., 2015. Optimization of wastewater microalgae saccharification using dilute acid hydrolysis for acetone, butanol, and ethanol fermentation. Appl. Energy 140, 14–19.
Chen, C., Wang, L., Xiao, G., Liu, Y., Xiao, Z., Deng, Q., et al., 2014. Continuous acetone-butanol-ethanol (ABE) fermentation and gas production under slight pressure in a membrane bioreactor. Bioresour. Technol. 163, 6–11.
Cheng, H.H., Whang, L.M., Chan, K.C., Chung, M.C., Wu, S.H., Liu, C.P., et al., 2015. Biological butanol production from microalgae-based biodiesel residues by *Clostridium acetobutylicum*. Bioresour. Technol. 184, 379–385.
dos Santos Vieira, C.F., Maugeri Filho, F., Maciel Filho, R., Pinto Mariano, A., 2019. Acetone-free biobutanol production: past and recent advances in the isopropanol-butanol-ethanol (IBE) fermentation. Bioresour. Technol. 287, 121425.
Dürre, P., 2011. Fermentative production of butanol—the academic perspective. Curr. Opin. Biotechnol. 22, 331–336.
Eckert, G., Schügerl, K., 1987. Continuous acetone-butanol production with direct product removal. Appl. Microbiol. Biotechnol. 27, 221–228.
Ellis, J.T., Hengge, N.N., Sims, R.C., Miller, C.D., 2012. Acetone, butanol, and ethanol production from wastewater algae. Bioresour. Technol. 111, 491–495.
Ennis, B.M., Marshall, C.T., Maddox, I.S., Paterson, A.H.J., 1986. Continuous product recovery by in-situ gas stripping/condensation during solvent production from whey permeate using *Clostridium acetobutylicum*. Biotechnol. Lett. 8, 725–730.
Ennis, B.M., Qureshi, N., Maddox, I.S., 1987. In-line toxic product removal during solvent production by continuous fermentation using immobilized *Clostridium acetobutylicum*. Enzyme Microb. Technol. 9, 672–675.
Errico, M., Sanchez-Ramirez, E., Quiroz-Ramìrez, J.J., Segovia-Hernandez, J.G., Rong, B.-G., 2016. Synthesis and design of new hybrid configurations for biobutanol purification. Comput. Chem. Eng. 84, 482–492.
Errico, M., Sanchez-Ramirez, E., Quiroz-Ramìrez, J.J., Rong, B.G., Segovia-Hernandez, J.G., 2017. Multiobjective optimal acetone-butanol-ethanol separation systems using liquid-liquid extraction-assisted divided wall columns. Ind. Eng. Chem. Res. 56, 11575–11583.
Evans, P.J., Wang, H.Y., 1988. Enhancement of butanol formation by *Clostridium acetobutylicum* in the presence of decanol-oleyl alcohol mixed extractants. Appl. Environ. Microbiol. 54, 1662–1667.
Ezeji, T.C., Qureshi, N., Blaschek, H.P., 2003. Production of acetone, butanol and ethanol by *Clostridium beijerinckii BA101* and in situ recovery by gas stripping. World J. Microbiol. Biotechnol. 19, 595–603.
Ezeji, T.C., Qureshi, N., Blaschek, H.P., 2004. Acetone butanol ethanol (ABE) production from concentrated substrate: reduction in substrate inhibition by fed-batch technique and product inhibition by gas stripping. Appl. Microbiol. Biotechnol. 63, 653–658.
Ezeji, T.C., Qureshi, N., Blaschek, H.P., 2007. Production of acetone butanol (AB) from liquefied corn starch, a commercial substrate, using *Clostridium beijerinckii* coupled with product recovery by gas stripping. J. Ind. Microbiol. Biotechnol. 34, 771–777.

Ezeji, T., Milne, C., Price, N.D., Blaschek, H.P., 2010. Achievements and perspectives to overcome the poor solvent resistance in acetone and butanol-producing microorganisms. Appl. Microbiol. Biotechnol. 85, 1697–1712.
Ezeji, T.C., Qureshi, N., Blaschek, H.P., 2013. Microbial production of a biofuel (acetone-butanol-ethanol) in a continuous bioreactor: Impact of bleed and simultaneous product removal. Bioprocess Biosyst. Eng. 36, 109–116.
Gottumukkala, L.D., Parameswaran, B., Valappil, S.K., Mathiyazhakan, K., Pandey, A., Sukumaran, R.K., 2013. Biobutanol production from rice straw by a non acetone producing *Clostridium sporogenes*. Bioresour. Technol. 145, 182–187.
Grisales-Díaz, V.H., Olivar-Tost, G., 2017. Energy efficiency of a new distillation process for isopropanol, butanol, and ethanol (IBE) dehydration. Chem. Eng. Process. Process Intensif. 112, 56–61.
Groot, W.J., Luyben, K.C.A.M., 1986. In situ product recovery by adsorption in the butanol/isopropanol batch fermentation. Appl. Microbiol. Biotechnol. 25, 29–31.
Groot, W.J., Van Der Lans, R.G.J.M., Luyben, K.C.A.M., 1992. Review technologies fermentations for butanol recovery integrated with fermentations. Process Biochem. 27, 61–75.
Huang, K., Wang, S.J., Shan, L., Zhu, Q., Qian, J., 2007. Seeking synergistic effect—a key principle in process intensification. Sep. Purif. Technol. 57, 111–120.
Huang, H., Liu, H., Gan, Y.R., 2010. Genetic modification of critical enzymes and involved genes in butanol biosynthesis from biomass. Biotechnol. Adv. 28, 651–657.
Hugo, V., Díaz, G., Tost, G.O., 2016. Butanol production from lignocellulose by simultaneous fermentation, saccharification, and pervaporation or vacuum evaporation. Bioresour. Technol. 218.
Ishii, S., Taya, M., Kobayashi, T., 1985. Production of butanol by *Clostridium acetobutylicum* in extractive fermentation system. J. Chem. Eng. Japan 18, 125–130.
Jang, Y.S., Lee, J., Malaviya, A., Seung, D.Y., Cho, J.H., Lee, S.Y., 2012. Butanol production from renewable biomass: rediscovery of metabolic pathways and metabolic engineering. Biotechnol. J. 7, 186–198.
Jin, C., Yao, M., Liu, H., Lee, C.F., Ji, J., 2011. Progress in the production and application of n-butanol as a biofuel. Renew. Sustain. Energy Rev. 15, 4080–4106.
Jones, D.T., Woods, D.R., 1986. Acetone-butanol fermentation revisited. Microbiol. Rev 50, 484–524.
Khanna, S., Goyal, A., Moholkar, V.S., 2013. Production of n-butanol from biodiesel derived crude glycerol using *Clostridium pasteurianum* immobilized on amberlite. Fuel 112, 557–561.
Komonkiat, I., Cheirsilp, B., 2013. Felled oil palm trunk as a renewable source for biobutanol production by *Clostridium* spp. Bioresour. Technol. 146, 200–207.
Larrayoz, M.A., Puigjaner, L., 1987. Study of butanol extraction through pervaporation in acetobutylic fermentation. Biotechnol. Bioeng. 30, 692–696.
Lee, S.H., Eom, M.H., Kim, S., Kwon, M.A., Choi, J.D.R., Kim, J., et al., 2015. Ex situ product recovery and strain engineering of *Clostridium acetobutylicum* for enhanced production of butanol. Process Biochem. 50, 1683–1691.
Li, J., Chen, X., Qi, B., Luo, J., Zhang, Y., Su, Y., et al., 2014. Efficient production of acetone-butanol-ethanol (ABE) from cassava by a fermentation-pervaporation coupled process. Bioresour. Technol. 169, 251–257.
Luyben, W.L., 2012. Pressure-swing distillation for minimum- and maximum-boiling homogeneous azeotropes. Ind. Eng. Chem. Res. 51, 10881–10886.
Mariano, A.P., Keshtkar, M.J., Atala, D.I.P., Filho, F.M., MacIel, M.R.W., Filho, R.M.I., et al., 2011. Energy requirements for butanol recovery using the flash fermentation technology. Energy and Fuels 25, 2347–2355.

Mariano, A.P., Filho, R.M., Ezeji, T.C., 2012a. Energy requirements during butanol production and in situ recovery by cyclic vacuum. Renew. Energy 47, 183–187.

Mariano, A.P., Qureshi, N., Maciel Filho, R., Ezeji, T.C., 2012b. Assessment of in situ butanol recovery by vacuum during acetone butanol ethanol (ABE) fermentation. J. Chem. Technol. Biotechnol. 87, 334–340.

Mariano, A.P., Dias, M.O.S., Junqueira, T.L., Cunha, M.P., Bonomi, A., Filho, R.M., 2013. Butanol production in a first-generation Brazilian sugarcane biorefinery: technical aspects and economics of greenfield projects. Bioresour. Technol. 135, 316–323.

Matsumura, M., Kataoka, H., Sueki, M., Araki, K., 1988. Energy saving effect of pervaporation using oleyl alcohol liquid membrane in butanol purification. Bioprocess Eng. 3, 93–100.

Merola, S.S., Tornatore, C., Marchitto, L., Valentino, G., Corcione, F.E., 2012. Experimental investigations of butanol-gasoline blends effects on the combustion process in a SI engine. Int. J. Energy Environ. Eng. 3, 1–14.

Ndaba, B., Chiyanzu, I., Marx, S., 2015. N-Butanol derived from biochemical and chemical routes: a review. Biotechnol. Reports 8, 1–9.

Nielsen, D.R., Prather, K.J., 2009. In situ product recovery of n-butanol using polymeric resins. Biotechnol. Bioeng. 102, 811–821.

Nielsen, L., Larsson, M., Holst, O., Mattiasson, B., 1988. Adsorbents for extractive bioconversion applied to the acetone-butanol fermentation. Appl. Microbiol. Biotechnol. 28, 335–339.

Outram, V., Lalander, C.A., Lee, J.G.M., Davies, E.T., Harvey, A.P., 2017. Applied in situ product recovery in ABE fermentation. Biotechnol. Prog. 33, 563–579.

Patraşcu, I., Bîldea, C.S., Kiss, A.A., 2017. Eco-efficient butanol separation in the ABE fermentation process. Sep. Purif. Technol. 177, 49–61.

Patraşcu, I., Bîldea, C.S., Kiss, A.A., 2018. Eco-efficient downstream processing of biobutanol by enhanced process intensification and integration. ACS Sustain. Chem. Eng. 6, 5452–5461.

Patraşcu, I., Bîldea, C.S., Kiss, A.A., 2019. Dynamics and control of a heat pump assisted azeotropic dividing-wall column for biobutanol purification. Chem. Eng. Res. Des. 146, 416–426.

Qian, Z., Chen, Q., Grossmann, I.E., 2018. Optimal synthesis of rotating packed bed and packed bed: a case illustrating the integration of PI and PSE. Computer Aided Chemical Engineering. Elsevier Masson SAS.

Quiroz-Ramírez, J.J., Sánchez-Ramírez, E., Segovia-Hernández, J.G., 2018. Energy, exergy and techno-economic analysis for biobutanol production: a multi-objective optimization approach based on economic and environmental criteria. Clean Technol. Environ. Policy 20, 1663–1684.

Qureshi, N., Hughes, S., Maddox, I.S., Cotta, M.A., 2005. Energy-efficient recovery of butanol from model solutions and fermentation broth by adsorption. Bioproc. Biosys. Eng. 27(4), 215–222.

Qureshi, N., Maddox, I.S., 1995. Continuous production of acetone-butanol-ethanol using immobilized cells of *Clostridium acetobutylicum* and integration with product removal by liquid-liquid extraction. J. Ferment. Bioeng. 80, 185–189.

Qureshi, N., Maddox, I.S., Friedl, A., 1992. Application of continuous substrate feeding to the ABE fermentation: relief of product inhibition using extraction, perstraction, stripping, and pervaporation. Biotechnol. Prog. 8, 382–390.

Rakopoulos, D.C., Rakopoulos, C.D., Giakoumis, E.G., Dimaratos, A.M., Kyritsis, D.C., 2010. Effects of butanol-diesel fuel blends on the performance and emissions of a high-speed di diesel engine. Energy Convers. Manag. 51, 1989–1997.

Remi, J.C. Saint, Baron, G., Denayer, J., 2012. Adsorptive separations for the recovery and purification of biobutanol. Adsorption 18, 367–373.

Roffler, S.R., Blanch, H.W., Wilke, C.R., 1987. In-situ recovery of butanol during fermentation - part 2: fed-batch extractive fermentation. Bioprocess Eng 2, 181–190.

Sánchez-Ramírez, E., Alcocer-García, H., Quiroz-Ramírez, J.J., Ramírez-Márquez, C., Segovia-Hernández, J.G., Hernández, S., et al., 2017a. Control properties of hybrid distillation processes for the separation of biobutanol. J. Chem. Technol. Biotechnol. 92, 959–970.

Sánchez-Ramírez, E., Quiroz-Ramírez, J.J., Hernández, S., Segovia-Hernández, J.G., Kiss, A.A., 2017b. Optimal hybrid separations for intensified downstream processing of biobutanol. Sep. Purif. Technol. 185, 149–159.

Van der Merwe, A.B., Cheng, H., Görgens, J.F., Knoetze, J.H., 2013. Comparison of energy efficiency and economics of process designs for biobutanol production from sugarcane molasses. Fuel 105, 451–458.

Van der Wal, H., Sperber, B.L.H.M., Houweling-Tan, B., Bakker, R.R.C., Brandenburg, W., López-Contreras, A.M., 2013. Production of acetone, butanol, and ethanol from biomass of the green seaweed *Ulva lactuca*. Bioresour. Technol. 128, 431–437.

van Wyk, S., van der Ham, A.G.J., Kersten, S.R.A., 2018. Pervaporative separation and intensification of downstream recovery of acetone-butanol-ethanol (ABE). Chem. Eng. Process. Process Intensif. 130, 148–159.

Wayman, M., Parekh, R., 1987. Production of acetone-butanol by extractive fermentation using dibutylphthalate as extractant. J. Ferment. Technol. 65, 295–300.

Weizmann, C., Bergmann, E., Sulzbacher, M., Pariser, E., 1948. Studies in selective extraction and adsorption III. The adsorption of acetone, butyl alcohol and 2,3-butylene glycol from dilute solutions. J. Soc. Chem. Ind. 53, 225–227.

Wiehn, M., Staggs, K., Wang, Y., Nielsen, D.R., 2014. In situ butanol recovery from *Clostridium acetobutylicum* fermentations by expanded bed adsorption. Biotechnol. Prog. 30, 68–78.

Xue, C., Zhao, J., Lu, C., Yang, S.T., Bai, F., Tang, I.C., 2012. High-titer n-butanol production by *Clostridium acetobutylicum* JB200 in fed-batch fermentation with intermittent gas stripping. Biotechnol. Bioeng. 109, 2746–2756.

Xue, C., Liu, F., Xu, M., Tang, I.C., Zhao, J., Bai, F., et al., 2016. Butanol production in acetone-butanol-ethanol fermentation with in situ product recovery by adsorption. Bioresour. Technol. 219, 158–168.

Yadav, S., Rawat, G., Tripathi, P., Saxena, R.K., 2014. A novel approach for biobutanol production by *Clostridium acetobutylicum* using glycerol: a low cost substrate. Renew. Energy 71, 37–42.

Yang, X., Tsai, G.J., Tsao, G.T., 1994. Enhancement of in situ adsorption on the acetone-butanol fermentation by *Clostridium acetobutylicum*. Sep. Technol. 4, 81–92.

Zhang, C., Li, T., He, J., 2018. Characterization and genome analysis of a butanol-isopropanol-producing *Clostridium beijerinckii* strain BGS1 06 Biological Sciences 0605 Microbiology 06 Biological Sciences 0604 Genetics. Biotechnol. Biofuels 11, 1–11.

Zheng, J., Tashiro, Y., Wang, Q., Sonomoto, K., 2015. Recent advances to improve fermentative butanol production: genetic engineering and fermentation technology. J. Biosci. Bioeng. 119, 1–9.

CHAPTER 6

Furfural

Contents

6.1 Introduction

The new environmental challenges bring out the need of replacing fossil resources with renewable alternatives, which it has encouraged researching improvements in the production of chemicals based on biomass. The biomass is formed mainly by three main fractions the cellulose, hemicellulose and lignin. The hemicellulose represents about 20%–40% of biomass content, which becomes it in the second most abundant polysaccharide in the biomass and nature (Saha, 2003). Hemicellulose is formed predominantly by heterogeneous chains of pentose polymers such as xylose and arabinose and to a lesser extent by chains of hexoses (glucose, mannose, and galactose) and sugar acids (Saha, 2003). Recently, the conversion and utilization of hemicellulose have received much interest owing to its possible portfolio of chemicals derived from it with several useful applications. However, the viability of the use of pentose sugars from hemicellulose, depends mostly on an efficient release and suitable conversion of these sugars. In this sense, furfural ($C_5H_4O_2$) is one of the most interesting products derived from hemicellulose sugars, because it is a natural degradation product of

Improvements in Bio-Based Building Blocks Production Through Process Intensification and Sustainability Concepts
DOI: https://doi.org/10.1016/B978-0-323-89870-6.00008-1

hemicellulose in an acid medium. Additionally, in recent years, the furfural has received special attention as a potential chemical platform for producing a wide portfolio of chemical products, for this reason, the furfural is also called "the sleeping beauty biorenewable chemical" (Marcotullio, 2011; Cai et al., 2014; Kabbour and Luque, 2020). A proof of this is that the United States National Renewable Energy Laboratory (NREL) and the European Commission of Energy have listed the furfural in the top of most promising chemicals produced from biomass, due to its many industrial applications and its wide variety of high valued derivatives (NREL, 2004; Taylor et al., 2015). Furfural is an aldehyde that at room temperature is a colorless liquid but turns yellow or brown in contact with air. As aforementioned, furfural is produced from pentoses contained in the hemicellulose, hence, biomasses with a high content of hemicelluloses and specially pentoses are remarkably preferable for producing furfural. Nowadays, the furfural production is based on the hydrolyzation of pentoses and their subsequent dehydration to furfural using dilute solutions of mineral acids at medium temperatures around 150-220°C. The most common acids used to produce furfural are sulfuric and chloride acid. A simplified reaction scheme to produce furfural from hemicellulose is shown in Fig. 6.1, similar reactions schemes have been reported by Mariscal et al. (2016).

Nowadays, the agricultural lignocellulosic wastes, especially the sugarcane bagasse and corn stover are the most used raw materials for furfural production, they comprise close to 98% of the total raw materials used in furfural production owing to their high availability, low costs, and high hemicellulose content in contrast with other feedstocks (Cai et al., 2014). Other common raw materials used to manufactured furfural and their respective hemicellulose content are reported in Table 6.1 these data were taken from the previous works reported by Cai et al. (2014) and Zeitsch (2000).

6.2 Uses of furfural

As already mentioned, the furfural has attracted attention because of its potential to produce a wide range of diverse products with different

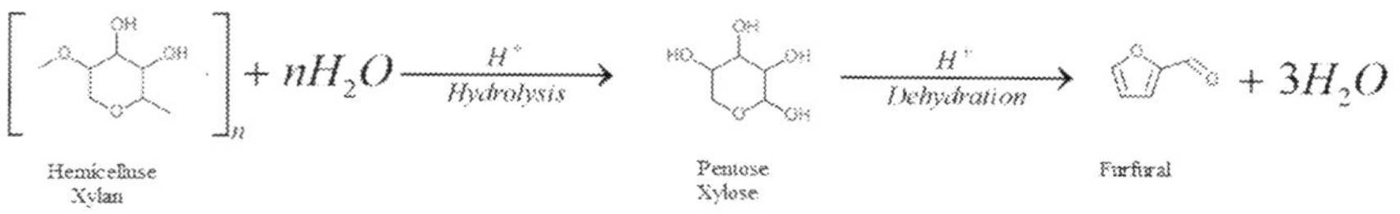

Figure 6.1 Furfural production route from hemicellulose.

Table 6.1 Main raw materials for furfural production.

Raw material	Hemicellulose content wt.% (dry basis)
Corn stover	30–35
Wheat straw	30–36
Rye straw	30
Oat hulls	29
Cottonseed hulls	28
Barley straw	28
Sugarcane bagasse	25–30
Sunflower husks	25
Hazelnut shells	23
Birchwood logs	22
Rice hulls	17
Maple wood	16
Pinewood	8
Peanut shells	3

applications in many downstream processes. As consequence, the number of uses for furfural is enormous. With this in mind, the furfural uses can be divided into two main groups, the current applications and the novel furfural applications. Currently, furfural is used as extract agent in the refining of lubricant oils, production of nematicides, fungicides and mainly to produce furfuryl alcohol (65%–70% of furfural global production or 5000 tons in 2003), which is used to produce resins (Cai et al., 2014; Hoydonckx et al., 2007; Zeitsch, 2000). In addition, during 50's the furfural was an important raw material in the production of Nylon 6–6 and other polymers as polyester, however, 1961 Du Pont abandoned this furfural-based path by a petrochemical route. Nevertheless, the route of nylon production from furfural has taken on a notorious importance again. Therefore, a inverse migration from oil-based routes in nylon production toward furfural-based routes is technically feasible without important modification to the current processes (Anthonia and Philip, 2015; Isikgor and Becer, 2015; Hoydonckx et al., 2007). Finally, in recent years several novel furfural applications have been proposed, such as: the production of thermal carbon composites to increase their resistance at high temperature in thermal shields (Pirolini, 2015), production of furan resins used as adhesives, sand agglomerator and cement resistant to chemical attacks (Anthonia and Philip, 2015) and the production of furonic acid, which is a very versatile chemical used widely in the pharmaceutical

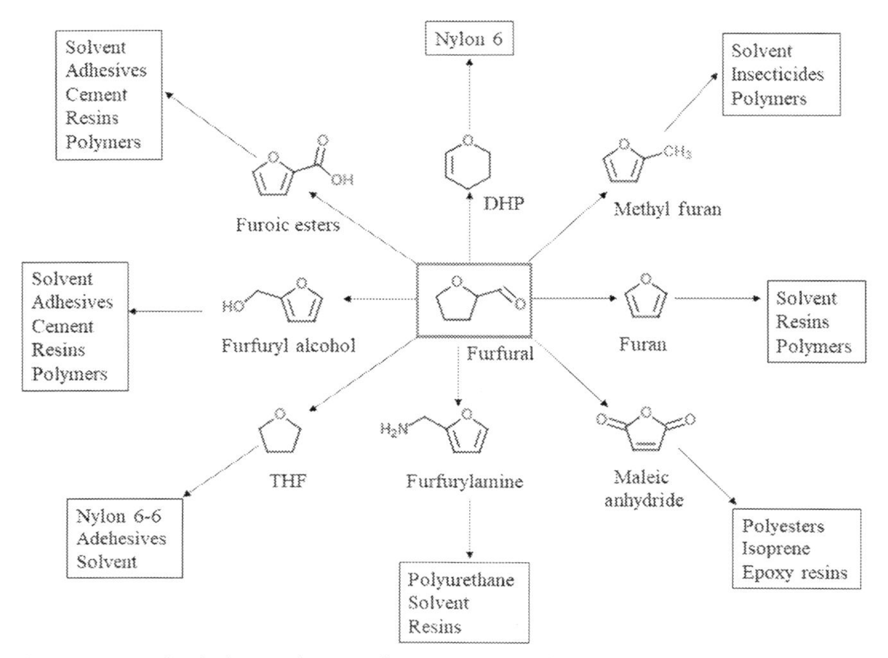

Figure 6.2 Furfural chemical routes for producing different commodities.

industry (Pirolini, 2015). The potential of furfural is enormous, nowadays the researchers are finding new applications and new furfural-based chemical routes for producing commodities, it is for this reason that the furfural market is continuously growing. Fig. 6.2 shows the main possible routes for producing different commodities from furfural and its applications.

6.3 Current furfural markets

Nowadays, furfural is used as intermediate chemical by several chemical industries, for this reason, it is difficult to estimate the demand for a specific application. However, data provided by Grand View Research (GVR) (2021) estimate that the furfural market has a value of USD 814.6 million and a demand of 461,380 tons, which gives a selling price of furfural about USD 1942 per ton. This data is according to estimates performed by other sources such as O'Brien (2006), which indicates that furfural demand in 2004 was estimated at around 300,000 tons with an annual increment of 2%; therefore, the demand for the year 2019 according to this estimate should be about 403,760 tons. In the same way,

Grand View Research (GVR), 2021 estimates that the demand for furfural will increase at a rate of 5% per year from 2019 to 2027. On the other hand, various sources confirm that the United States and Europe are two of the main consumers of furfural. Fig. 6.3 shows some important furfural markets and their demand. These data were taken from O'Brien (2006) and actualized to the year 2020 using his annual growth estimation of 2%.

With respect to the principal furfural producers, China is the main furfural supplier, in 2001 it supplied about 70% of world demand (Kabbour and Luque, 2020). In addition, some sources have determined that for the year 2020 China provided around 80%–85% of global demand (Kubic et al., 2020; GVR, 2021). China has 200 furfural production plants distributed in all the country, with an average production of 1000 tons/year, where the company Tieling North Furfural Group is the major China's producer with a production of 50,000 tons/year of furfural (O'Brien, 2006; Kabbour and Luque, 2020). On the other hand, the Central Romana Corporation located in Dominican Republic is considered the major furfural facility with production of about 11% of world supplies. In Spain, the Furfural Español Corporation has a furfural production capacity of 5000 tons/year, whereas in India the company KRBL, the major supplier has a production of 10,000 tons/year where with a production of 3000 tons/year (Kabbour and Luque, 2020). Fig. 6.4 shows the furfural production level for different countries.

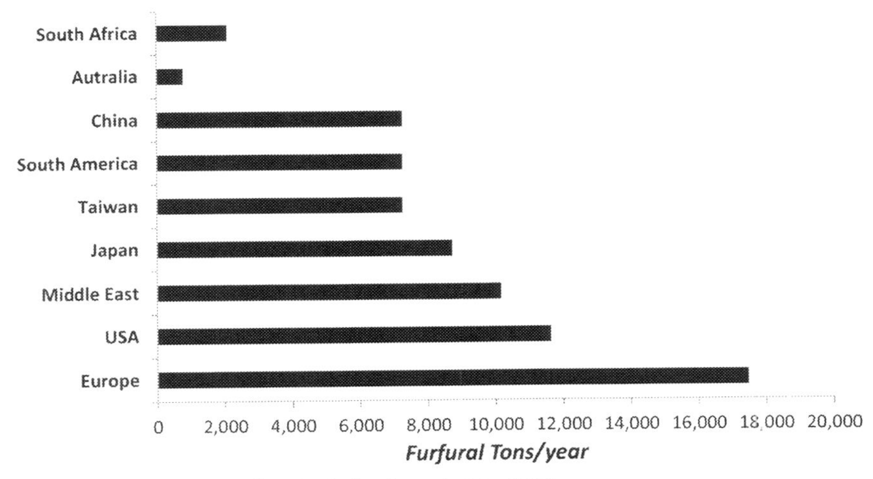

Figure 6.3 Estimation of main furfural market in 2020.

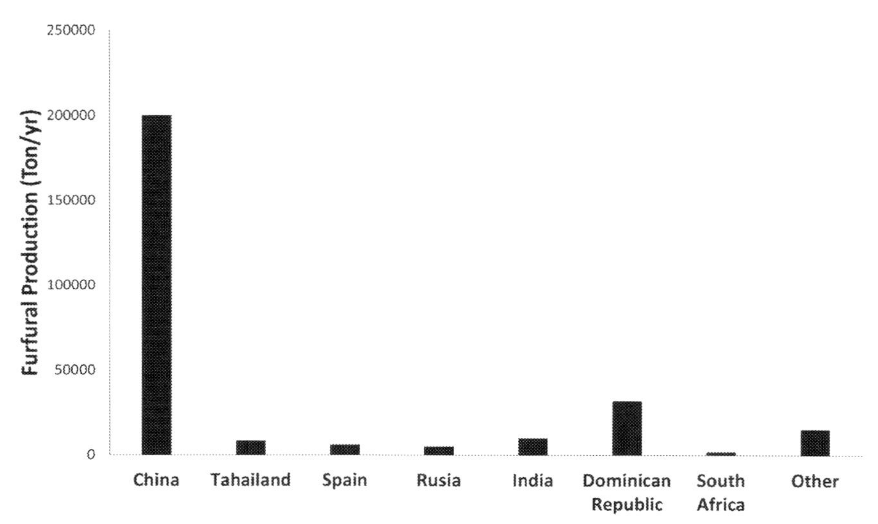

Figure 6.4 Main furfural producers. *Data taken from Kabbour, M., Luque, R., 2020. Chapter 10 — Furfural as a platform chemical: From production to applications. In: S. Saravanamurugan, A. Pandey, H. Li, A. Riisager (Eds.), Biomass, Biofuels, Biochemicals, 283—297. Elsevier. https://doi.org/10.1016/B978-0-444-64307-0.00010-X.*

6.4 Stoichiometric and kinetics models for furfural production

In order to understand the production of furfural from biomass, it is of the utmost importance to understand the stoichiometry and analyze the underlying kinetics. As mentioned earlier, furfural is primarily produced by the degradation of hemicellulose in dilute aqueous solutions of mineral acids. Therefore, the kinetics and stoichiometric presented in this chapter are based on the reaction of hemicellulose in acid solutions. The previous work reported by Zeitsch (2000) is considered as of the most important references, in this work Zeitsch (2000) reported the stoichiometry of furfural production from hemicellulose in a single process, which can be expressed as follows:

$$\begin{array}{ccc} \text{Xylan} - 2\cdot\text{Water} & \rightarrow & \text{Furfural} \\ C_5H_8O_4 - 2\cdot H_2O & \rightarrow & C_5H_4O_2 \\ 132.114 - 36.032 & \rightarrow & 96.082 \end{array} \tag{6.1}$$

Based on the stoichiometry, the maximum theoretical yield can be calculated as follows:

$$Y = \frac{96.082}{132.114} = 0.727 \tag{6.2}$$

This theoretical yield is of utmost importance because it represents a measure of efficiency for the furfural production processes.

Concerning the chemical kinetics to furfural from hemicellulose or xylose, it is important to highlight, that unlike cellulose, hemicellulose is not chemically homogeneous from one raw material to another. The composition of hemicellulose varies enormously depending on the type of biomass, for example, the hemicellulose contained in hardwood is constituted mainly by xylans, while the softwood hemicelluloses contain glucomannans, for this reason, the hardwood is preferred to produce furfural, whereas the softwoods are ideal for producing other chemicals such as levulinic acid (McMillan, 1994; Saha, 2003). This variability on hemicellulose composition for different raw materials has as a consequence, that there is not exist a unique kinetic model to express the furfural formation, but rather the kinetics of furfural formation depends strongly on biomass used. Therefore, a specific kinetic model and its respective parameters should be determined for a specific biomass. Despite this limitation, researchers have tried to propose general chemical routes to try to express the kinetics of furfural formation for a large number of biomasses, changing only the kinetic parameters from one feedstock to another. As result of this, one of the most extended chemical mechanism used to express the kinetic models in furfural production is the Saeman model (Saeman, 1945), which has been used in several different works and biomasses with good results (Bustos et al., 2003; Aguilar et al., 2002; Bamufleh et al., 2013; Li et al., 2019). The generalized Seaman model can be expressed according to the following elementary reaction series:

$$\text{Polymers} \rightarrow \text{Monomers} \rightarrow \text{Descomposition Products} \tag{6.3}$$

This generalized Seaman model applies for the degradation of any polymer such as xylan, glucan, glucomannans, and so on. In the case of furfural production, the Seaman model must be applied to the xylan fraction in order to calculate the furfural produced. Therefore, the Seaman model can be rewritten according to the following elementary reactions:

$$\text{Xylan} \xrightarrow{k_1} \text{Furfural} \xrightarrow{k_2} \text{Descomposition Products} \tag{6.4}$$

Where k_1 and k_2 are the specific reaction rates. On the other hand, the decomposition products involve different componds such as, resins, methanol and acetic acid formed during the reaction. Commonly, the specific reaction rates in the Seaman Model can be calculated using a modified Arrhenius equation to determine the effect of acid concentration (C)

(Aguilar et al., 2002; Bamufleh et al., 2013). In this sense, the specific reaction rates (k_n) are calculated as follows:

$$k_1 = aC^m \exp\left(-\frac{E}{RT}\right) \tag{6.5}$$

Where a is the preexponential constant, m is a regression parameter and C is the acid concentration in (wt. %/wt. %). Some kinetic parameters for Seamen model (Eqs. 6.4 and 6.5) are shown in Table 6.2, those kinetic parameters were determined for different biomasses and acid catalysts.

Finally, it is important to mention that although the Seaman model is the most common model used to determinate the furfural formation, it is not the only kinetic model reported, thus a suitable kinetic model should be sought for a specific catalyst and biomass, some examples of this are kinetic models reported by Marcotullio (2011), Lavarack et al. (2002) and Bamufleh et al. (2013).

6.5 Current technologies for furfural production

The furfural production is not new, it is one of the oldest renewable chemicals produced by the human, owing to it is a coproduct of fermentation during ethanol production (Peters, 1936). The first record about a selective production of furfural date from the year 1831, when Dobereiner isolate furfural for first time, afterward Fownes in 1845 produced furfural through degradation of bran with mineral acids. For this reason, Fownes named furfural to this component, said name comes from the Latin root furfur = bran. Nevertheless, the first industrial process to produce furfural date from 1922, it was developed by the company Quaker Oats (Zeitsch, 2000). Typically, this process uses oat hulls, corn cobs, and sugar cane bagasse as raw materials for producing furfural. In this process, the biomass is fed jointly a sulfuric acid solution (2.2 wt.% on dry biomass basis) to a degradation reactor. In this reactor, the biomass is treated at the thermal conditions of 153°C for 5 hour, under these conditions the energy consumption is close to 42.9 GJ per metric ton of furfural (Marcotullio, 2011). This energy is supplied by a vapor stream which is fed to the degradation reactor at a rate of 16.2 kg of steam per kg of furfural produced, under these conditions the theoretical energy required for the production of the saturated steam is around 2.64 MJ/kg. This saturated vapor stream is of the utmost importance because it has the function of providing energy to the reactor and removing the furfural as quickly as possible in

Table 6.2 Seaman model parameters for furfural production.

Raw material/ catalyst	Specific reaction rate (k_1)	Specific reaction rate (k_2)	References
Palm tree/H_2SO_4	$k_1 = (6.24 \times 10^8/min)C^{1.2}\exp\left(-\frac{93.94KJ/moL}{RT}\right)$	$k_2 = (1.74/\text{min})C^{2.06}\exp\left(-\frac{33.63KJ/moL}{RT}\right)$	Romero-García et al. (2020)
Eucalyptus sawdust/H-SAPO-34	$k_1 = (7.6 \times 10^9/min)C^{1.19}\exp\left(-\frac{120.41KJ/moL}{RT}\right)$	Not reported	Li et al. (2019)
Sugarcane bagasse/ H_2SO_4	$k_1 = (1.0 \times 10^8/s)C^{0.7217}\exp\left(-\frac{103.1KJ/moL}{RT}\right)$	$k_2 = (4.12 \times 10^4/s)C^{0.7217}\exp\left(-\frac{60.3KJ/moL}{RT}\right)$	Lavarack et al. (2002)
Sugarcane bagasse/ HCl	$k_1 = (87.2 \times 10^{15}/s)C^{1.87}\exp\left(-\frac{165.6KJ/moL}{RT}\right)$	$k_2 = (1.28 \times 10^{14}/s)C^{1.87}\exp\left(-\frac{131.1KJ/moL}{RT}\right)$	Lavarack et al. (2002)
Corn stover/ H_2SO_4	$k_1 = (1.4 \times 10^{14}/min)C^{0.68}\exp\left(-\frac{111.6KJ/moL}{RT}\right)$	$k_1 = (1.4 \times 10^{14}/min)C^{0.68}\exp\left(-\frac{111.6KJ/moL}{RT}\right)$	Jin et al. (2011)

order to avoid its resinification (Zeitsch, 2000; Marcotullio, 2011). Furfural leaves from the degradation reactor with an average concentration of 5.8 wt.% where the water is the main compound, so once the furfural has been produced, the outlet steam from said reactor is condensed and sent to a distillation process where the furfural is purified. A general scheme for Quaker oats process is presented in Fig. 6.5.

Despite the excessive energy consumption required by the Quaker Oats process, this process is considered as one of the most important advances in furfural production, even now it is used to produce around 80% of the worldwide furfural production, due to its easy implementation in contrast to other alternatives. Additionally, it is important to highlight, that this process has not undergone important improvements from 1922 (Zeitsch, 2000; Marcotullio, 2011; Nhien et al., 2017). Owing to this process requires intensive use of energy, water, and it is characterized by excessive biomass degradation, which impossibilities the use of other biomass fractions to produce other biochemicals, in the last two decades, the advances in the furfural field have been focused on how to increase the profitability and reduce the environmental impact and energy consumption (Taylor et al., 2015; Nhien et al., 2016).

Biofine process is a good example of advances in furfural production, this process was invented by Fitzpatrick in 1990 (De Jong and Marcotullio, 2010). The Biofine process tries to increase the profitability through coproduction of furfural and levulinic acid, a biochemical

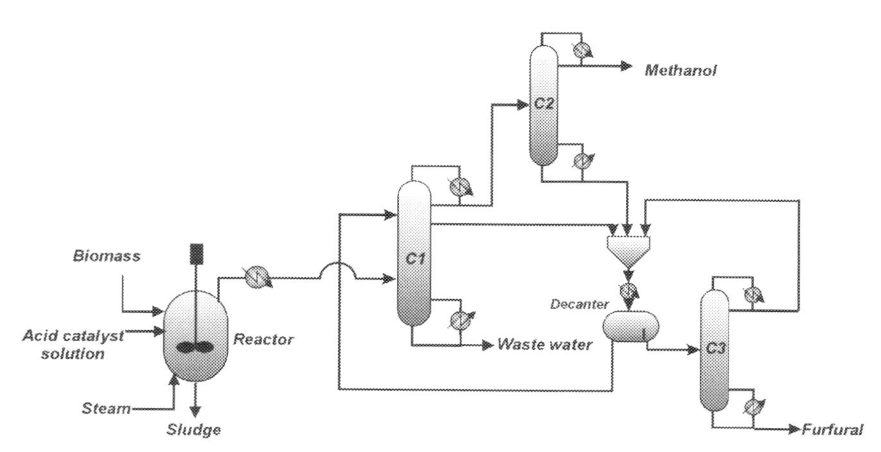

Figure 6.5 Quaker Oats process. *Adapted from Zeitsch, K.J., 2000. The Chemistry and Technology of Furfural and Its Many By-Products, vol. 13. Elsevier.*

produced from the cellulose fraction of biomass. A process scheme of Biofine process is shown in Fig. 6.6. This process consists in two reactors, first the biomass is mixed with a dilute solution of sulfuric acid, this solution has a concentration around 1.5–3 wt.% of sulfuric acid. Then the biomass and sulfuric acid is sent to a plug flow reactor which works at the thermal conditions of 210°C–220°C and short residence time in order to avoid the furfural resignification, and a excessive cellulose degradation. Subsequently, the hydrolyzed biomass is fed to a CSTR reactor, in this reactor the xylose and glucose sugars are transformed into furfural and levulinic acid, respectively. This second reactor works at temperatures about 190°C–200°C, pressure of 14 bar with a residence time of 20 minutes. These conditions benefit the production of levulinic and not furfural, because cellulose is more difficult to hydrolyze and dehydrate in contrast to hemicellulose, for this reason, the Biofine process is preferred to produce levulinic acid instead furfural.

Another important furfural production process is the Verdernikos process, which uses small aliquots of strong acids to perform the degradation on pentosans (xylans) of hemicellulose into furfural. This process addresses two important problems in furfural production, poor yields and cellulose degradation (Vedernikov, 1996). A scheme of Vedernikov's process is illustrated in Fig. 6.7. The Vedernikov's process consists of treating the biomass with small aliquots of dilute acid solutions, then the biomass mixed with the sulfuric acid is sent to a hydrolysis reactor where, another vapor stream is fed in order to remove the furfural as quickly as possible from the acid medium avoinding the furfural resinification and excessive

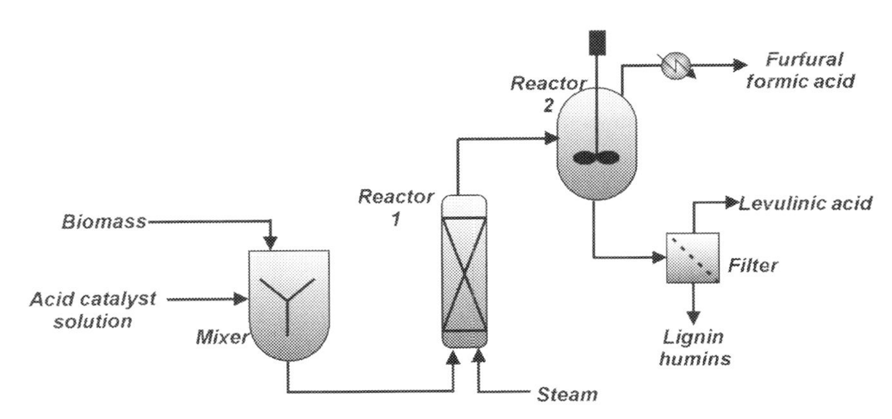

Figure 6.6 Biofine process.

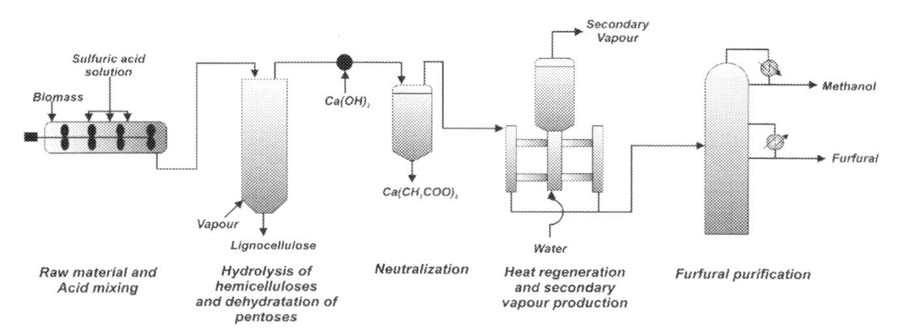

Figure 6.7 Vedernikov process (Vedernikov, 1996).

degradation of cellulose. Afterward, the organic acids and the sulfuric acid have been neutralized with a solution of calcium hydroxide. The stream rich in furfural and some products is condensed and mixed with water, in this step, the residual vapor which has not been condensed is recovered and reused. Finally, the mixture of furfural, coproducts is fed to a distillation column where they are purified.

Ultimately, it is important to mention that there are other advances in furfural production, such as the Suprayield process proposed by Zeitsch (2000) or the MTC process (O'Brien, 2006). Nevertheless, all those new advances have been reported on pilot plant scale, thus their profitability has not been proved at an industrial scale. Finally, Table 6.3 summarizes and compares some important data for different industrial furfural processes.

6.6 New intensified proposes for furfural production

Despite the multiple advances on furfural production, some important problems persist or have not been fully solved, such as the high operating and investment costs, high energy and water consumption, inefficient preservation and utilization of other biomass fractions to produce different chemicals are some of the main challenges that should be addressed in order to improve the feasibility of furfural production (Cai et al., 2014; De Jong and Marcotullio, 2010).

6.6.1 Advances in furfural purification

Process intensification and process optimization have showed to be important tools that can improve the current furfural production process or help

Table 6.3 Comparison of different furfural production technologies.

Process	Catalyst	Raw material (Biomass)	Coproducts	Furfural yield (% theoretical)	References
Quaker Oats	H_2SO_4	Oat hulls/bagasse	Methanol, acetic acid, resins	50	Zeitsch (2000)
Biofine	H_2SO_4	Agricultural residues/paper sludge	Levulnic acid, formic acid, lignin	70	De Jong and Marcotullio (2010)
Vedernikov	H_2SO_4	Wood chips	Methanol, resins, acetic acid	75	De Jong and Marcotullio (2010), O'Brien (2006), Vedernikov (1996), Gravitis et al. (2001)
MTC	H_2SO_4	Straw	Cellulose, HMF	85	Marcotullio (2011), De Jong and Marcotullio (2010)
Melbourne University	–	–	Phenol, cellulose	65	O'Brien (2006)
Huaxia	H_2SO_4	Corn cobs	Methanol, acetic acid, acetone, levulinic acid	35	O'Brien (2006)
Suprayield	H_2SO_4	–	–	72	Zeitsch (2000)

in the development and design of new furfural processes. In this way, new advances and studies in furfural production have been reported in several journals, in recent years. Many of the studies have been focused in improved the profitability and reduce the environmental impact of the Quaker Oats process. The reason for proposing to improve the Quaker process instead of designing new alternatives is mainly since this process is used to produce about 80% of the global production of furfural, due to its easy implementation. With this in mind, one of the stages of the Quaker Oats process that is most feasible and easily improved by intensification is the separation stage.

Two of the most representative works on furfural purification published in recent years are the works reported by Nhien et al. (2016) and Contreras-Zarazúa et al. (2018). Nhien et al. (2016) proposed different intensified separation processes, such as dividing wall columns and heat integration with the aim of reducing the total annual cost (TAC) and the emission in the furfural purification stage. Contreras-Zarazúa et al. (2018) studied and optimized the columns proposed by Nhien et al. (2016) and also, they proposed two additional thermally coupled systems. Fig. 6.8 shows the distillation schemes studied by Contreras-Zarazúa et al. (2018), the processes consist in the Quaker Oats process, which was used as a benchmark. Additionally, a thermally coupled configuration (TCC), a thermally equivalent configuration (TEC), a dividing wall column configuration (DWC) and heat integrated option (HIC) were considered. These separation processes were designed and optimized considering a feed composition of water 90 wt.%, furfural 6 wt.%, methanol 2 wt.%, and acetic acid 2 wt.%, this composition was selected because it corresponds to the average composition produced by several furfural reactors (Zeitsch, 2000). On the order hand, the mass flowrate was 105,000 kg/hour, which is the mass flow rate required for satisfying the furfural demand (Nhien et al., 2016). Besides, it is important to mention, that these processes were simultaneously designed and optimized using the differential evolution with Tabu list (DETL) method, which was proposed by Srinivas and Rangaiah (2007). The DETL method is a stochastic global search optimization technique, which emulates the biological evolution process. More information about this optimization method is provided by several previous works such as Srinivas and Rangaiah (2007) and Rangaiah (2010). The optimization method was implemented in a hybrid platform that links Excel through Visual Basic with Aspen Plus. In this case, the optimization method was programmed in Visual Basic within Excel, whereas in

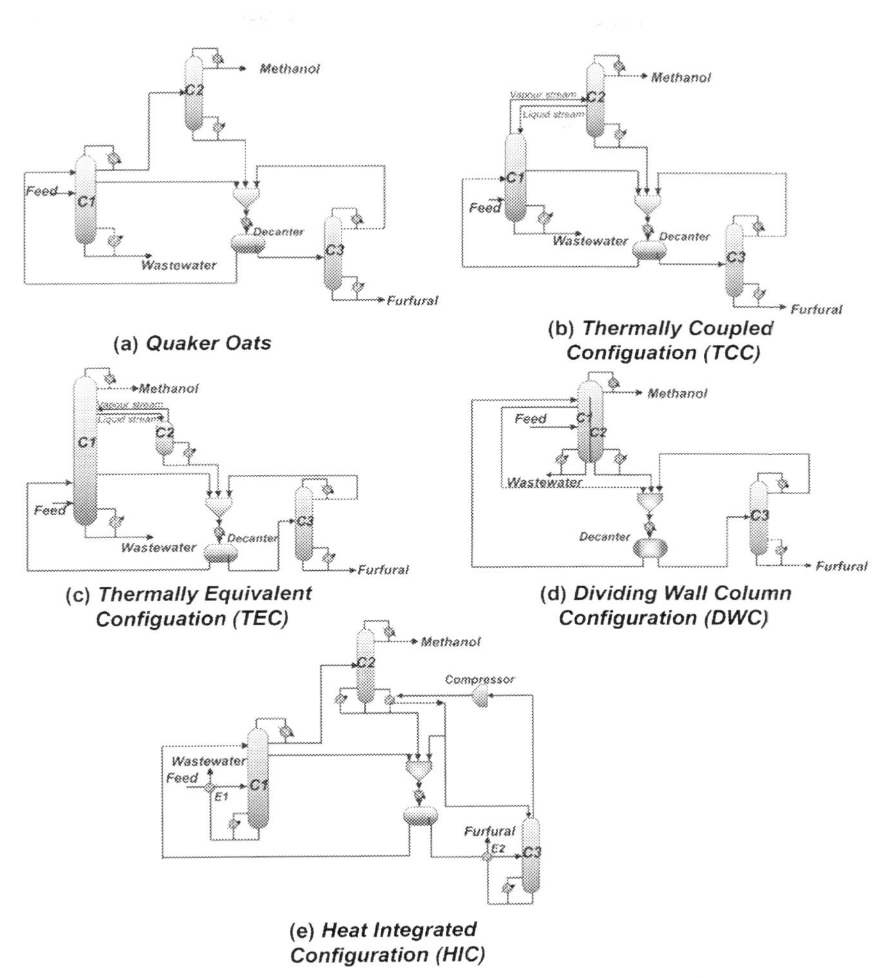

Figure 6.8 Distillation sequences studied by Contreras-Zarazúa et al. (2018).

Aspen Plus is performed the rigorous simulations of possible solutions proposed by the DETL method.

The objective functions used by Contreras-Zarazúa et al. (2018) to evaluate the performance of the separation processes are the Total Annual Cost (TAC), Eco-Indicator 99 (EI99) and Individual Risk (IR), these metrics evaluate the economic, environmental and safety aspects of the processes, respectively. These objective functions are described in the following subsection.

6.6.2 Objective functions

TAC was the metric used to evaluate the economics of the processes. The TAC consists of the sum of annualized investment costs also called capital cost and the utilities costs. The TAC can be expressed mathematically according to the next equation:

$$\text{TAC} = \frac{\text{Capital cost}}{\text{Payback period}} + \text{Operating cost} \tag{6.6}$$

The investment costs are the expenditures associated with condensers, reboilers, distillation columns, trays, process vessels and compressors. On the other hand, the utilities costs are associated with the expenditures of electricity, cooling water, steam, solvents, and so on. The Guthrie methodology was used to calculate the investment costs. The equation and parameters used by Contreras-Zarazúa et al. (2018) for estimating both capital and utilities costs were taken from Turton et al. (2008). Carbon steel was used as construction material, whereas a payback period of 10 years was considered.

In the case of the environmental impact function, it is evaluated using EI99 methodology, which is life cycle assessment method. The EI99 was proposed by Goedkoop and Spriensma (2001) and it has been used to for evaluating several chemical processes (Contreras-Zarazúa et al., 2020). The EI99 is based on the evaluation of eleven impact categories grouped into three main subcategories: ecosystem quality, resources depletion and human health. The EI99 provides results in eco points where one eco-point symbolizes the 1,000th part of the environmental load generated by an average European inhabitant. For distillation columns and other process separations the most important factors that increase the environmental impact are the steel required to build the process equipment, the electricity required for pumping cooling water and the steam used for heating. Mathematically, the EI99 can be calculated according to the follow equation:

$$\text{EI99} = \sum_{b} \sum_{d} \sum_{k \in K} \delta_d \omega_d \beta_b \alpha_{b,k} \tag{6.7}$$

Where β_b is the amount of chemical b released by direct emission per unit of, $\alpha_{b,k}$ represents the damage in category k per unit of chemical b released to the environment, ω_d is the damage weighting factor of category d, and δ_d is a normalization factor for damage of category d, respectively. Contreras-Zarazúa et al. (2018) proposed to evaluate the EI99

Table 6.4 Impact category values for EI99 (Goedkoop and Spriensma, 2001).

Impact category	Steel (points/kg) × 10^{-3}	Steam (points/kg)	Electricity (points/kWh)
Carcinogenic	1.29×10^{-3}	1.180×10^{-4}	4.360×10^{-4}
Climate change	1.31×10^{-2}	1.27×10^{-3}	4.07×10^{-3}
Ionizing radiation	4.510×10^{-4}	1.91×10^{-6}	8.94×10^{-5}
Ozone depletion	4.550×10^{-6}	7.78×10^{-7}	5.41×10^{-7}
Respiratory effects	8.010×10^{-2}	1.56×10^{-3}	1.01×10^{-5}
Acidification	2.710×10^{-3}	1.21×10^{-4}	9.88×10^{-4}
Ecotoxicity	7.450×10^{-2}	2.85×10^{-4}	2.14×10^{-4}
Land occupation	3.730×10^{-3}	8.60×10^{-5}	4.64×10^{-4}
Fossil fuels	5.930×10^{-2}	1.24×10^{-2}	1.01×10^{-2}
Mineral extraction	7.420×10^{-2}	8.87×10^{-6}	5.85×10^{-5}

considering a hierarchical perspective with the aim to obtain an equilibrium among short- and long-term effects. For this reason, the damages to human health and to the ecosystem are considered with equal importance, thus both are equally weighted, while damage to resources is half as important, which corresponds to a typical hierarchical perspective (Goedkoop and Spriensma, 2001). The impact parameters using by Contreras-Zarazúa et al. (2018) to evaluate the EI99 are illustrated in Table 6.4.

Finally, the IR was the index considered for evaluating safety evaluation.

The IR quantifies the probability of injury or death to a person located at a specified distance from the epicenter of an accident. This index is independent of the number of people exposed to a hazard (Center for Chemical Process Safety (CCPS), 2000). The mathematical expression for IR is shown in Eq. 6.8.

$$\mathrm{IR} = \sum f_i P_{x,y} \tag{6.8}$$

Where f_i is the frequency of occurrence of specific incident i, and $P_{x,y}$ is the likelihood of affectation produced by the incident i. The quantitative risk analysis (QRA) was the methodology used to identify and quantify the potential incidents and their respective consequences. The QRA method classifies the incidents into two main categories which are continuous and instantaneous releases. A continuous release is originated owing to a partial rupture on a pipeline or process vessel generating a leak of material. On other hand, an instantaneous release is provoked due to a catastrophic rupture of a process equipment which results in the total loss of matter from the process vessel. More information about this method and the possible incidents is provided by CCPS (2000). Based on QRA methodology, jet fire, flash fire and toxic releases were identified as continuous release incidents, whereas the instantaneous incidents determined are boiling liquid expanding vapor explosion, unconfined vapor cloud explosion, flash fire and toxic releases.

Once the possible incidents have been identified, the probability of affectation ($P_{x,y}$) can be calculated through a consequence analysis assessment, which consists of determining physical variables such as; the thermal radiation, overpressure and leak concentration generated during an incident and their effects (damage) on a person located at a specific distance from this incident. The equations required for calculating those physical variables are reported by CCPS (2000) and Medina-Herrera et al. (2014). Once the physical variables have been calculated, the damage is calculated using a probit model, which relatesthe values of physical variables to damage produced to a person. Commonly, the damage considered to evaluate the probit model is the death of a person. The probit equations that correlate the decease owing to thermal radiation ($t_e E_r$) and overpressure are shown as follows:

$$Y = -14.9 + 2.56 \ln\left(\frac{t_e E_r^{\frac{4}{3}}}{10^4}\right) \tag{6.9}$$

$$Y = -77.1 + 6.91 \ln(p^{\circ}) \tag{6.10}$$

These equations were taken from CCPS (2000). The calculations were performed contemplating a representative distance of 50 m. Finally, once the probit variable (Y) has been calculated, the probability affectation ($P_{x,y}$) is determined as follows:

$$P_{x,y} = 0.5\left[1 + erf\left(\frac{Y-5}{\sqrt{2}}\right)\right] \tag{6.11}$$

It is important to highlight, that for toxic realizes the median lethal concentration (LC50) was used to compute the affectation probability, because there are not probit models reported for all the chemical

Table 6.5 Physical properties of components (NIOSH, 2020).

Component	Lower flammability limit (LFL)	Upper flammability limit (UFL)	Median lethal concentration (LC50)	Heat combustion (kJ/moL)
Furfural	2	19	64,000 ppm/4 h	2344
Methanol	6	36	1037 ppm/1 h	726
Acetic acid	6	17	16,000 ppm/4 h	876.1

substances associated with the furfural production. Finally, an atmospheric stability type F with a wind speed of 1.5 m/second is considered to compute the concentrations of chemical substances in air due to toxic releases. This condition corresponds to the worst scenario possible (Crowl and Louvar, 2001). The physical properties of the compounds implicated in the furfural purification are reported in Table 6.5.

Once the economic, environmental and safety objective functions have been explained, the mathematical formulation of optimization problem can be performed as follows:

The mathematical optimization problem considering all indexes can be expressed according to:

$$\begin{gathered} \min\ [TAC, EI99] = f(NT_i, ANT_i, Fs_i, R_i, VF, LF, DC_i, HD_i) \\ \text{Subject to:} \\ \vec{y}_m \geq \vec{x}_m \\ \vec{w}_m \geq \vec{u}_m \end{gathered} \tag{6.12}$$

where NT_i is the number of stages of column i, Fs_i is the feed stage for the distillation column i, Ri is the reflux ratio of column i, VF is the interconnecting vapor flow, LF is the interconnecting liquid flow, DCi is the diameter of column i, HDi is the reboiler duty for the distillation column i. The optimization was performed in order to satisfy the constraint vectors of purity and mass flowrate for the furfural and methanol, respectively. The purity requirements for both components are 99.5%–99.2% for furfural and methanol respectively. The parameters used for the DETL method are: number of population (NP): 120 individuals, generations number (GenMax): 710, Tabu list size: 60 individuals, Tabu radius: 0.01, crossover fractions (Cr): 0.8, mutation fractions (F): 0.3. Finally, the decision variables considered for each distillation system are shown in Table 6.6.

Table 6.6 Decision variables for intensified Quaker Oats processes.

Decision variables	Quaker Oats	TCC	TEC	DWC	HIC
Number of stages for columns	X	X	X	X	X
Feed stage of recycle	X	X	X	X	X
Stage of side stream C1	X	X	X	X	X
Main feed stage	X	X	X	X	X
Mass flow side stream C1	X	X	X	X	X
Reflux ratio	X	–	X	–	X
Heat duty	X	X	X	X	X
Diameters of columns	X	X	X	X	X
Discharge pressure of compressor	–	–	–	–	X
Interlinking flow	–	X	X	X	–
Heat integrated	–	–	–	–	X
Total number of variables	18	18	18	18	21

6.6.3 Optimization results

In this section the optimization results are presented. In this case, the Pareto front method is used to illustrate the results, while the utopian point methodology is used for selecting the solution with the best trade-off between the different objectives. Fig. 6.9 shows the Pareto fronts for all distillation processes. Remember that for this study the conversional Quaker Oats process was used as a benchmark.

Fig. 6.9A shows the Pareto front the eco-indicator versus TAC. It is important to highlight that each dot of Pareto front symbolizes a design for a specific separation scheme. Those Pareto Charts were obtained after 85,200 iterations, which results of multiplying the 710 generations and 120 individuals. Therefore, those points correspond to the results of last generation. The shape of Pareto front indicates that the EI99 and TAC follow the same trend since both indexes are strongly influenced by the energy required in these processes. This energy is supplied to the processes as steam, which is generated by the combustion of fuels such as natural gas, thus more energy implies an increase of fuel burned, which increases the environmental impact. On the other hand, more fuel required means that the utilities costs will also increase. In this chart, the further to the right the points are located, the processes equipment are larger, with more stages and more energy consumption.

Fig. 6.9B corresponds to the Pareto front of IR versus TAC. Note, that in this case the objectives have an antagonist behavior, as a

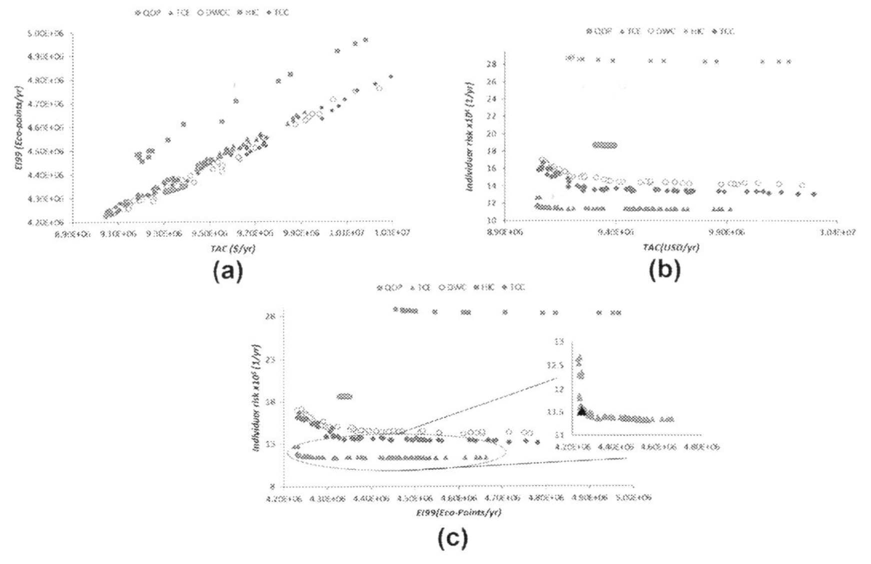

Figure 6.9 Pareto front for the intensified Quaker Oats separation processes (Contreras-Zarazúa et al., 2018).

consequence when an index improves the other one will deteriorate. This result can be easily explained. The safety objective quantifies some physical variables such as concentration of a toxic release or thermal radiation emitted during an incident. Those physical variables depend on the concentration of organic compounds inside a process vessel as result a higher concentration of those compounds within the process vessel implies that the incident is more lethal (CCPS, 2000). For this specific case, the water is the most abundant compound in the mixture, this water is removed mainly in the first column. In the case of design with better safety properties, not all the water is removed in this first column. Therefore, an important amount of the water ends up inside C2 and C3, which dilutes the concentration of organic compounds within these equipment. However, the separation costs are affected because more stages and energy are required to achieve the separation. This increment in the energy required also affects the environmental impact, for this reason when the safety index is improved the environmental impact is affected too (See Fig. 6.9C).

Finally, please note that the intensified processes cannot obtain important improvements on environmental impact and costs in relation to the conventional Quaker Oast process. These results can be explained by analyzing the composition of the mixture. Note that the water represents the 90 wt.% of the mixture, this water is separated in column C1, therefore, it is reasonable to consider that most of the energy consumed takes place in this column. As a consequence, the energy consumption cannot be decreased significantly, if the composition of the water at the entace to the distillation sequence is not reduced first. Consequently, the selection of the best separation process should be made based on the safety index, which presents significant changes. Based on this criterion, the TEC sequence has the lowest risk in contrast to other sequences, because it separates the methanol and water in the same column, which reduces significantly the risk of concentrating methanol in another column. The selection of the best design was performed using the utopic point, which is represented in the Fig. 6.9C. this point has the best trade-off among IR and EI99, owing to the same trend that the TAC this point also has the best trade-off among risk and costs. The selected design as best design is shown in Fig. 6.10, please note, that column C1 has about 95% of the

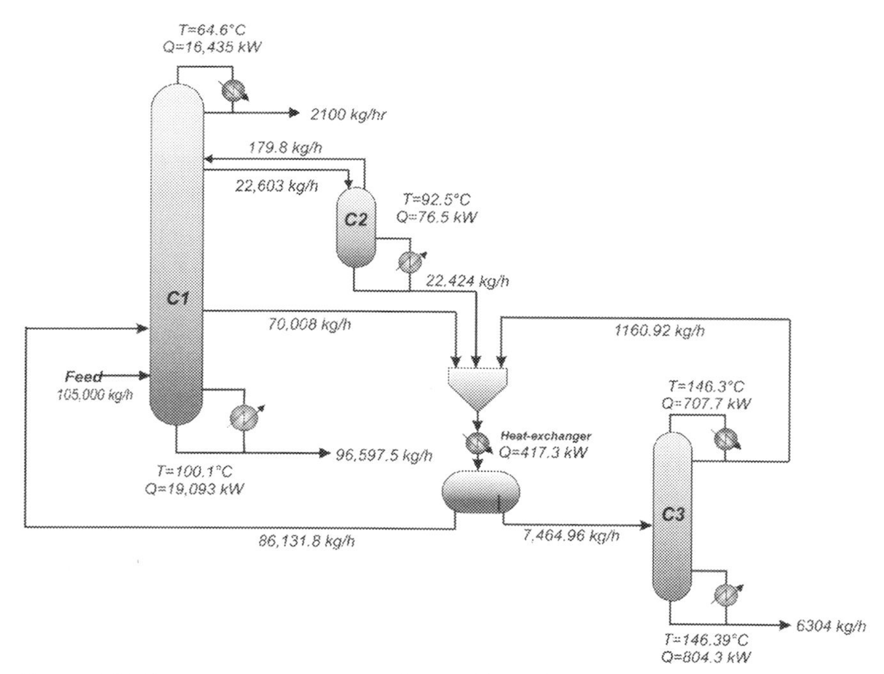

Figure 6.10 Energy requirements and mass flows for the selected equivalent thermally process.

total energy requirements, which indicates that the purification of water is the factor that most affects the energy consumption. Consequently, the conclusion is that the amount of water that is fed to the separation process must be decreased in order to improve the profitability and environmental impact of furfural separation, this point will be addressed in the following section.

6.6.4 Advances in furfural purification using hybrid extractive distillation schemes

In the last sections, some intensified schemes for the Quaker Oats process were analyzed and optimized under economic, environmental and safety criteria. The conclusion of the previous works reported by Nhien et al. (2016) and (Contreras-Zarazúa et al., 2018) was that it is necessary to decrease the water composition in the feed stream in order to improve the profitability and environmental impact in the furfural separation stage. In this way, one of the most promising alternatives to reduce the water composition is the liquid-liquid extraction.

A liquid-liquid extraction process consists of adding an entrainer substance also called solvent, which has the purpose of removing one of the components from the mixture. This entrainer must have some special features, however the most important is that the solvent cannot generate an azeotrope with the component to be removed, in order to not increase the complexity of separation. In this sense, Nhien et al. (2017) performed the search of a solvent capable to separate the water or the furfural from the mixture. However, owing to the complexity of the mixture the search is not trivial. They found that Butyl chloride is the most promising solvent to remove the furfural from the mixture, based on this solvent they proposed a hybrid extractive distillation process to purify furfural, which is shown in Fig. 6.11. Note, that in order to perform a representative comparison among this process and the Quaker Oast alternatives presented in their previous work (Nhien et al., 2016), the same feed composition and flow were considered by Nhien et al. (2017). In this work is concluded that important energy savings can be achieved in contrast to the traditional Quaker Oats alternatives, at the same time, it is concluded that due to the low composition of methanol and acetic acid the purification of these compounds is not economically feasible. On the other hand, Contreras-Zarazúa et al. (2020) have suggested that the methanol purification in the scheme proposed by Nhien et al. (2017) is feasible when suitable intensified processes are implanted. Therefore, a

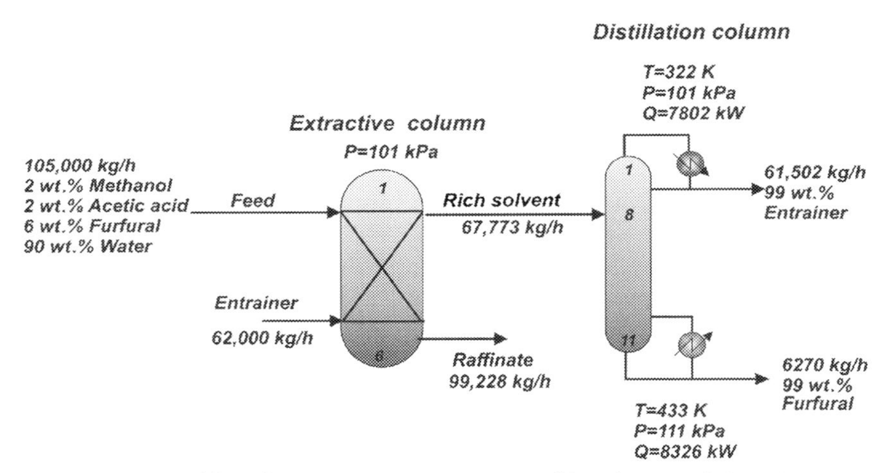

Figure 6.11 Liquid-liquid extraction process proposed by Nhien et al. (2017).

suitable methanol purification will increase significantly the profitability of the process. Based on this hypothesis, Contreras-Zarazúa et al. (2020) proposed different alternatives to purify the methanol from the bottom of the extractive column, those options are illustrated in Fig. 6.12.

It is important to mention, that Contreras-Zarazúa et al. (2020) modified the design proposed by Nhien et al. (2017) in order to decrease the amount of solvent required. This modification was performed through a sensitivity analysis. Fig. 6.13A shows the sensitivity analysis results and the final specifications for extractive column and distillation column C1. The sensitivity analysis shows how the TAC can decrease by reducing the solvent amount. In this Figure, each dot represents one design for a sequence of extraction column and a distillation column. The minimum TAC corresponds to a solvent mass flow of 47,000 kg/hour. For lower solvent amounts, the separation of the mixture is infeasible. The design with the minimum solvent amount is shown in Fig. 6.13B.

Once the sensitivity analysis has been completed, the design of methanol purification schemes is performed. The purities considered for designing the distillation schemes are, 99.6%–97.9% for methanol and water respectively. In the work of (Contreras-Zarazúa et al., 2020) different indexes were analyzed to evaluate and compare the different options. These indexes are the TAC, EI99, individual risk and energy required. these metrics were explained in the previous sections. The moving section methodology was used to design the intensified options and the

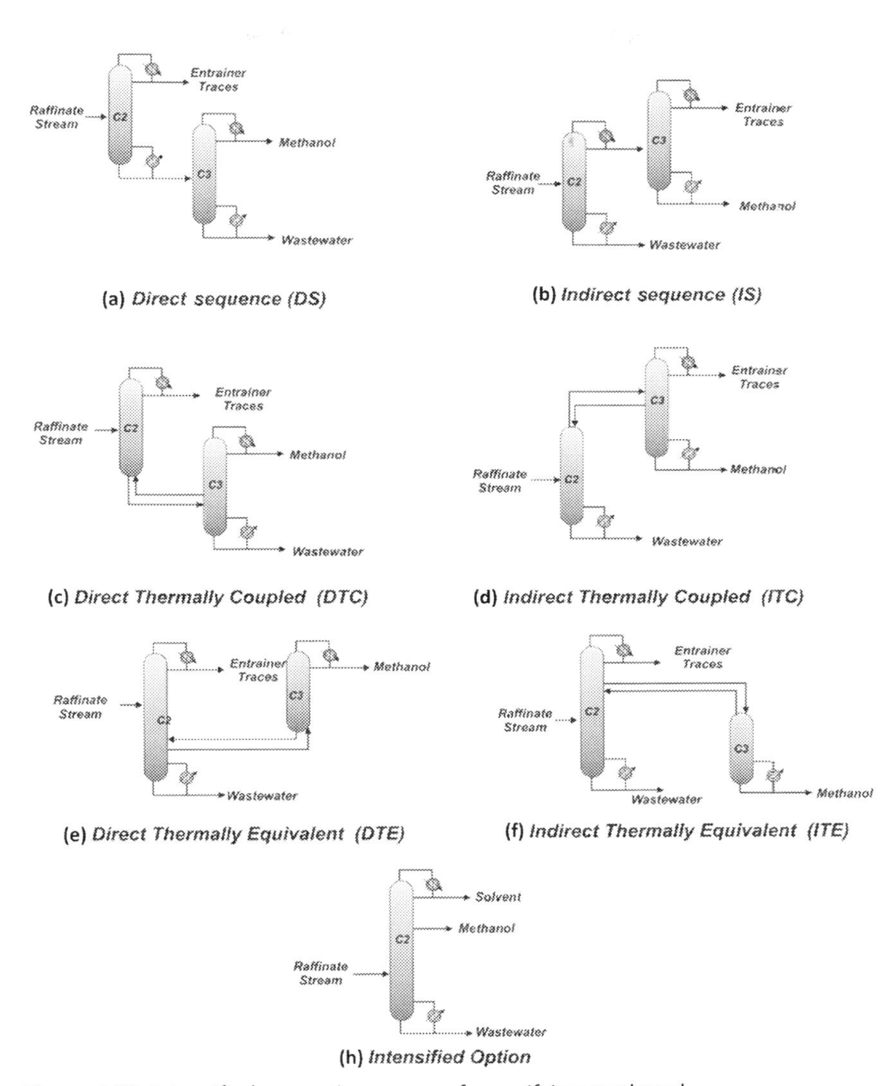

Figure 6.12 Intensified separation process for purifying methanol.

interconnecting mass flows where varied in order to determine the minimum energy consumption according to the methodology proposed by Hernández and Jiménez (1996). The results obtained for the sensitivity analysis of the interconnecting mass flows and the evaluation of the different indexes are presented in Fig. 6.14. Please note that the intensified option is not presented in Fig. 6.14A, because it does not have any interconnecting mass flow.

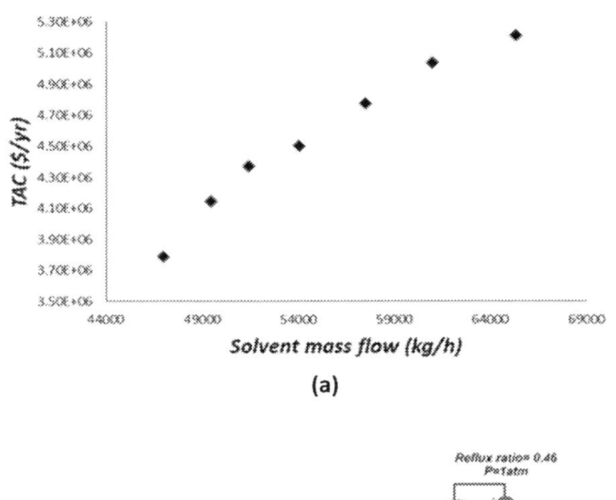

(a)

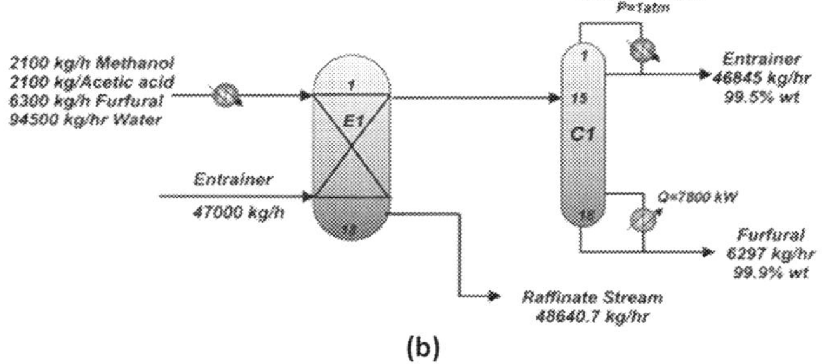

(b)

Figure 6.13 (A) Sensitivity analysis (B) modified hybrid extractive distillation design.

Note that the conventional sequences (DS and IS) have the worst values for all the metrics, whereas the intensified alternatives present important improvements. In this case, the IS, it shows the lowest value for all metrics. The reason is quite simple, the IS performs the methanol purification in only one equipment, thus important improvements on energy, TAC and environmental impact can be achieved. On the other hand, the safety index depends on the number of process equipment, fewer columns signify less probability of failure and therefore less probability of an accident, which improves the safety index. Additionally, note that in the IS option the methanol purification is performed in a single column with a larger amount of water, which generated a low concentration of methanol within the column and contributes to a better safety index. Based on the overall results, it is evident that the intensified sequence is the best option to purify methanol, thus a process scheme with the energy requirements and mass flows is illustrated in Fig. 6.15A,

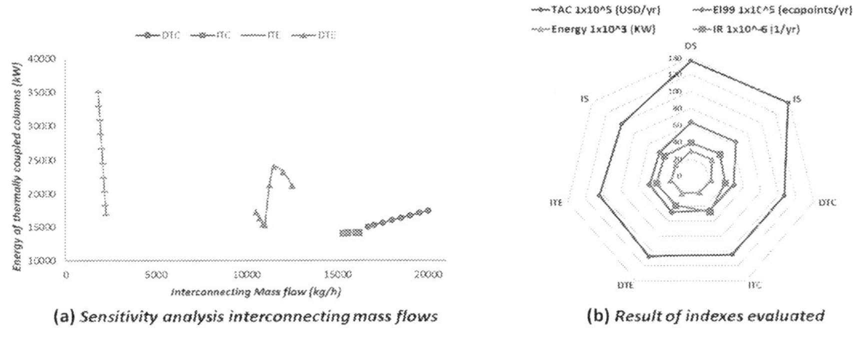

(a) *Sensitivity analysis interconnecting mass flows* (b) *Result of indexes evaluated*

Figure 6.14 (A) Sensitivity analysis (B) modified hybrid extractive distillation design.

more information about this sequence the other processes are provided by Contreras-Zarazúa et al. (2020). Based on Fig. 6.15A, please note that the purification of methanol represents the most important contribution to energy requirements in contrast to furfural purification. The reason is related to the amount of water that must be removed in order to purify the methanol. The large quantity of water implies that important energy requirements are required for separating the methanol. Therefore, thus like in the Quaker Oats alternative, the quantity of water from the reactor is the main factor that contributes to the energy consumption, costs, and environmental impact. In order to determine if the purification of methanol is economically feasible, the profit by methanol sales is compared with the purification costs of furfural and methanol (see Fig. 6.15B). The methanol's sale price considered for this analysis was 850 USD/ton, it was obtained from Alibaba Web (2020). Please note, that the profits due to methanol sales exceed the purification costs, thus, it is possible to support the process separation for methanol and furfural using only the methanol sales. Therefore, the profits due to furfural's sales can be considered as net profits, for those reasons the methanol recovery is economically feasible. In contrast to the intensified alternatives for the Quaker Oats process, the alternatives base on liquid-liquid extraction have similar costs.

6.7 Conclusions

In this chapter, some important aspects about furfural one of the most promisors' chemicals produced from biomass were revised. The furfural is a chemical produced from hemicellulose fraction, it has a wide range of application and a well establish current market and demand. Nowadays,

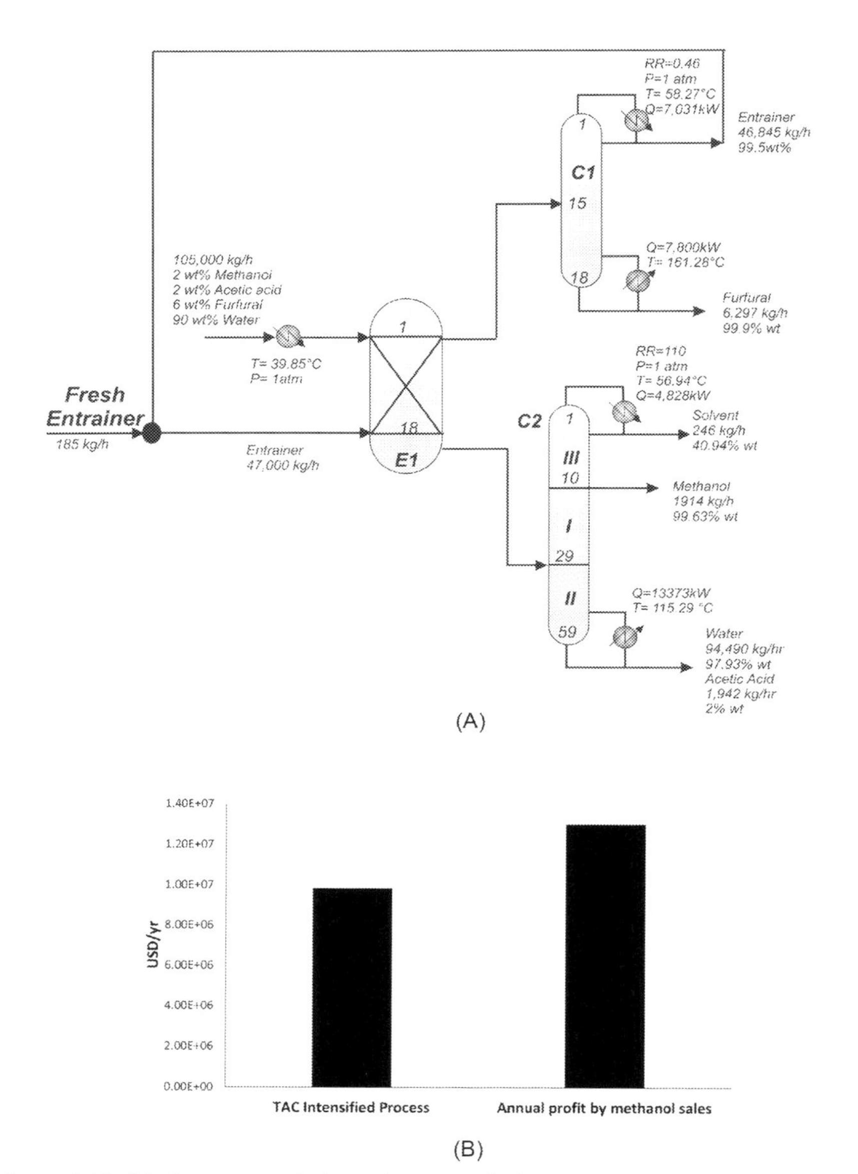

Figure 6.15 (A) Mass energy balance for intensified process (B) comparison of costs for intensified process and methanol sales.

exists different opportunities and alternatives to overcome some challenges in furfural production, some examples of this are the Verdernikovs and the biofine processes. This chapter was focused on improvements and advances in the Quaker Oast process, which is the most common furfural production process. In this sense, two different options to purify furfural were revised, one alternative based on azeotropic distillation and the other one based on liquid-liquid extraction. The results for both processes are clear, the purification of methanol is important in order to the profitability of the process. On the other hand, it can be concluded that advances and improvements in the reaction stage are required in order to increase the profitability of the process. It is important to highlight, that these improvements must be focused on two main points, the first one is the reduction of water required in the reaction stage and the second one consists of avoiding excessive furfural resignification.

References

Aguilar, R., Ramırez, J.A., Garrote, G., Vázquez, M., 2002. Kinetic study of the acid hydrolysis of sugar cane bagasse. J. food Eng. 55 (4), 309–318.

Alibaba Web, 2020. Alibaba Web, 2029. https://www.alibaba.com/product-detail/99-Purity-Methanol-methyl-Alcohol-Supplier_62513604124.html?spm = a2700.galleryofferlist.0.0.3ff266724j4c7T. (accessed 03 2020).

Anthonia, E.E., Philip, H.S., 2015. An overview of the applications of furfural and its derivatives. Int. J. Adv. Chem. 3, 42–47.

Bamufleh, H.S., Alhamed, Y.A., Daous, M.A., 2013. Furfural from midribs of date-palm trees by sulfuric acid hydrolysis. Ind. Crop. Prod. 42, 421–428.

Bustos, G., Ramírez, J.A., Garrote, G., Vázquez, M., 2003. Modeling of the hydrolysis of sugar cane bagasse with hydrochloric acid. Appl. Biochem. Biotechnol. 104 (1), 51–68.

Cai, C.M., Zhang, T., Kumar, R., Wyman, C.E., 2014. Integrated furfural production as a renewable fuel and chemical platform from lignocellulosic biomass. J. Chem. Technol. Biotechnol. 89 (1), 2–10.

Center for Chemical Process Safety (CCPS), 2000. Guidelines for Chemical Process Quantitative Risk Analysis. Center for Chemical Process Safety of the American Institute of Chemical Engineers.

Contreras-Zarazúa, G., Sánchez-Ramírez, E., Vazquez-Castillo, J.A., Ponce-Ortega, J.M., Errico, M., Kiss, A.A., et al., 2018. Inherently safer design and optimization of intensified separation processes for furfural production. Ind. Eng. Chem. Res. 58 (15), 6105–6120.

Contreras-Zarazúa, G., Jasso-Villegas, M.E., Ramírez-Márquez, C., Sánchez-Ramírez, E., Vázquez-Castillo, J.A., Segovia-Hernandez, J.G., 2020. Design and intensification of distillation processes for furfural and co-products purification considering economic, environmental, safety and control issues. Chem. Eng. Process 159, 108218.

Crowl, D.A., Louvar, J.F., 2001. Chemical Process Safety: Fundamentals with Applications. Pearson Education.

De Jong, W., Marcotullio, G., 2010. Overview of biorefineries based on co-production of furfural, existing concepts and novel developments. Int. J. Chem. React. Eng. 8 (1). Available from: https://doi.org/10.2202/1542-6580.2174.

Goedkoop, M., Spriensma, R., 2001. The eco-indicator 99, a damage oriented method for life cycle impact assessment, methodology report. PRé Consultants BV, 132 pp.

Grand View Research (GVR), 2021, https://www.grandviewresearch.com/industry-analysis/furfural-market.

Gravitis, J., Vedernikov, N., Zandersons, J., Kokorevics, A., 2001. Furfural and levoglucosan production from deciduous wood and agricultural wastes. ACS Symp. Series 784, 110–122.

Hernández, S., Jiménez, A., 1996. Design of optimal thermally-coupled distillation systems using a dynamic model. Chem. Eng. Res. Des. 74 (3), 357–362.

Hoydonckx, H.E., Van Rhijn, W.M., Van Rhijn, W., De Vos, D.E., Jacobs, P.A., 2007. Furfural and derivatives. Ullmann's Encyclopedia of Industrial Chemistry. Wiley-VCH Verlag GmbH & Co. KgaA. Available from: https://doi.org/10.1002/14356007.a12_119.pub2.

Isikgor, F.H., Becer, C.R., 2015. Lignocellulosic biomass: a sustainable platform for the production of bio-based chemicals and polymers. Polym. Chem. 6, 4497–4559.

Jin, Q., Zhang, H., Yan, L., Qu, L., Huang, H., 2011. Kinetic characterization for hemicellulose hydrolysis of corn stover in a dilute acid cycle spray flow-through reactor at moderate conditions. Biomass Bioenergy 35 (10), 4158–4164.

Kabbour, M., Luque, R., 2020. Chapter 10 – Furfural as a platform chemical: From production to applications. In: Saravanamurugan, S., Pandey, A., Li, H., Riisager, A. (Eds.), Biomass, Biofuels, Biochemicals. Elsevier, pp. 283–297. Available from: https://doi.org/10.1016/B978-0-444-64307-0.00010-X.

Kubic, W.L., Yang, X., Moore, C.M., Sutton, A., 2020. A Process for Converting Corn Bran to Furfural without Mineral Acids (No. LA-UR-20–24035). Los Alamos National Lab. (LANL), Los Alamos, NM.

Lavarack, B.P., Griffin, G.J., Rodman, D., 2002. The acid hydrolysis of sugarcane bagasse hemicellulose to produce xylose, arabinose, glucose and other products. Biomass Bioenergy 23 (5), 367–380.

Li, X., Yang, J., Xu, R., Lu, L., Kong, F., Liang, M., et al., 2019. Kinetic study of furfural production from Eucalyptus sawdust using H-SAPO-34 as solid Brønsted acid and Lewis acid catalysts in biomass-derived solvents. Ind. Crop. Prod. 135, 196–205.

Marcotullio, G., 2011. The chemistry and technology of furfural production in modern lignocellulose-feedstock biorefineries. Doctor of Philosophy, Delft University of Technology, L'Aquila, Italy.

Mariscal, R., Maireles-Torres, P., Ojeda, M., Sádaba, I., Granados, M.L., 2016. Furfural: a renewable and versatile platform molecule for the synthesis of chemicals and fuels. Energy Environ. Sci. 9 (4), 1144–1189.

McMillan, J.D., 1994. Pretreatment of lignocellulosic biomass. In: Himmel, M.E., Baker, J.O., Overend, R.P. (Eds.), Enzymatic Conversion of Biomass for Fuels Production. American Chemical Society, Washington, DC, pp. 292–324.

Medina-Herrera, N., Jiménez-Gutiérrez, A., Mannan, M.S., 2014. Development of inherently safer distillation systems. J. Loss. Prev. Process Ind. 29, 225–239.

National Institute for Occupational Safety and Health (NIOSH), 2020. https://www.cdc.gov/niosh/index.htm.

National Renewable Energy Laboratory (NREL), 2004. https://www.nrel.gov/docs/fy04osti/35523.pdf.

Nhien, L.C., Long, N.V.D., Kim, S., Lee, M., 2016. Design and optimization of intensified biorefinery process for furfural production through a systematic procedure. Biochem. Eng. J. 116, 166–175.

Nhien, L.C., Long, N.V.D., Kim, S., Lee, M., 2017. Techno-economic assessment of hybrid extraction and distillation processes for furfural production from lignocellulosic biomass. Biotechnol. Biofuels 10 (1), 81.

O'Brien, P., 2006. Furfural Chemicals and Biofuels from Agriculture. Rural Industries Research and Development Corporation, Kingston, p. 39.

Peters Jr, F.N., 1936. The furans: fifteen years of progress. Ind. Eng. Chem. 28 (7), 755–759.
Pirolini, A.P., 2015. Materials used in space shuttle thermal protection systems. Available from: https://www.azom.com. (accessed 20.07.19).
Rangaiah, G.P. (Ed.), 2010. Stochastic Global Optimization: Techniques And Applications In Chemical Engineering (With CD-ROM), vol. 2. World Scientific.
Romero-García, A.G., Prado-Rúbio, O.A., Contreras-Zarazúa, G., Ramírez-Márquez, C., Ramírez-Prado, J.H., Segovia-Hernández, J.G., 2020. Simultaneous design and controllability optimization for the reaction zone for furfural bioproduction. Ind. Eng. Chem. Res. 59 (36), 15990–16003.
Saeman, J.F., 1945. Kinetics of wood saccharification-hydrolysis of cellulose and decomposition of sugars in dilute acid at high temperature. Ind. Eng. Chem. 37 (1), 43–52.
Saha, B.C., 2003. Hemicellulose bioconversion. J. Ind. Microbiol. Biotechnol. 30 (5), 279–291.
Srinivas, M., Rangaiah, G.P., 2007. Differential evolution with Tabu list for solving nonlinear and mixed-integer nonlinear programming problems. Ind. Eng. Chem. Res. 46 (22), 7126–7135.
Taylor, R., Nattrass, L., Alberts, G., Robson, P., Chudziak, C., Bauen, A., et al., 2015. From the sugar platform to biofuels and biochemicals: final report for the European Commission Directorate-General Energy. E4tech/Re-CORD/Wageningen UR.
Turton, R., Bailie, R.C., Whiting, W.B., Shaeiwitz, J.A., 2008. Analysis, Synthesis and Design of Chemical Processes. Pearson Education.
Vedernikov, N., 1996. Process for producing of furfural. Patent LV11032, Letvia.
Zeitsch, K.J., 2000. The Chemistry and Technology of Furfural and Its Many By-Products, vol. 13. Elsevier.

CHAPTER 7

Levulinic acid

Contents

7.1 Introduction

The transition toward a more sustainable and green economy depends strongly on the feasible replacement of current raw materials by suitable renewable chemicals. For this reason, in recent years the efforts have been focused on efficient leverage and conversion of biomass, especially lignocellulosic residues into high value chemicals. In this sense, cellulose has attracted attention as a potential source of green chemicals due to its high availability. Cellulose is the most abundant fraction of biomass and represents about 35%−55% of the total content of lignocellulosic biomass (Mukherjee et al., 2015; Pileidis and Titirici, 2016; Kang et al., 2018). The global availability of cellulose has been estimated around 720 billion tons whereas 40 billion tons are naturally restored each year (Morone et al., 2015; Van de Vyver et al., 2011). The conversion of cellulose into chemicals is not new, a good example is the production of ethanol, which has been carried out for centuries. Currently, one of the most promising cellulose-derived chemicals is levulinic acid (LA), which has been listed by the United States National Renewable Energy Laboratory (NREL) in the top of 12 biochemicals derived from biomass (NREL, 2004). At the same time, the European Commission of Energy considers LA to be a key compound produced from lignocellulosic residues owing

Improvements in Bio-Based Building Blocks Production Through Process Intensification and Sustainability Concepts
DOI: https://doi.org/10.1016/B978-0-323-89870-6.00009-3

to a wide number of different applications (Werpy and Petersen, 2004; Taylor et al., 2015; Kang et al., 2018).

LA, also called gamma ketovaleric acid or 4-oxopentanoic acid, is a C5 chemical compound, classified within short chain fatty acid with chemical formula $C_5H_8O_3$. At room temperature, LA is a yellow or brown crystalline solid., it has a density of $1.1335 g/cm^3$ and a melting and boiling point of 33°C-245°C, respectively (Morone et al., 2015; ChemicalBook, 2021). LA is manufactured in a similar way to furfural, it is produced mainly by the acid degradation of cellulose fraction, therefore biomasses with high contents of cellulose are preferred to manufacture LA. Additionally, LA can be produced from hemicellulose fraction through of furfural to produce furfuryl alcohol and its subsequent dehydration into LA (Adeleye et al., 2019; Rackemann and Doherty, 2011). The chemical pathways for LA formation from cellulose and hemicellulose are shown in Fig. 7.1.

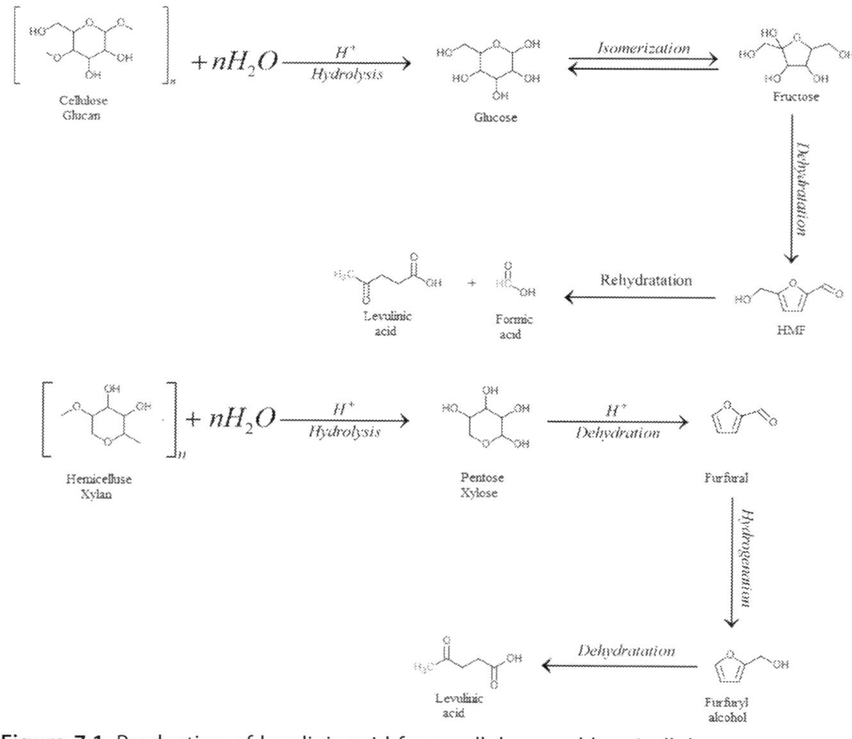

Figure 7.1 Production of levulinic acid from cellulose and hemicellulose.

In an analogous way to furfural production, the reactions presented in Fig. 7.1 are carried out under solutions of mineral acids at temperatures from 100° to 250°, and where the acid concentration plays an key role in order to achieve an efficient conversion of biomass into LA. Usually acid concentrations around 1–5%wt. are required to achieve the LA production (Bozell, et al., 2000; Signoretto et al., 2019; Fang and Qi, 2017). The most commons mineral acids used during levulinic production are hydrochloride and sulfuric acid (Girisuta, 2007).

As aforementioned, the biomasses rich in celluloses are preferable for producing LA. In this sense, the lignocellulosic residues especially the agricultural wastes are the most common feedstocks to produce LA, owing to their low costs (Adeleye et al., 2019). Table 7.1 shows the most common raw materials used in LA production and their respective cellulose content. the information was taken from the previous work reported by (Rackemann and Doherty, 2011) and Galletti et al. (2012).

7.2 Current uses of levulinic acid

Nowadays, LA stands out over other chemicals derived from cellulose, due to its several applications and derivatives, which becomes the LA into a versatile and useful building block. The versatility of LA is provided because, it has two different functional groups, which makes it easier to react with many chemicals. Some important examples of LA applications are the production of γ valerolactone which is used as solvent an eco-friendly fuel. Additionally, the LA is used for manufacturing ethyl levulinate an important intermediate chemical commonly used with

Table 7.1 Common raw materials for levulinic acid production.

Raw material	Cellulose content wt.% (dry basis)
Paper	85
Pulp residues	80
Sorghum grain	73.8%
Poplar sawdust	57.6
Paper sludge	57.1
Wheat straw	40.4%
Sugarcane bagasse	42%
Corn stover	39.6%
Tobacco cops	25%
Water hyacinth	26.3%

pharmaceutical purposes. Nevertheless, the production of polymers and plastics is the area where the LA stans out, it can be used for producing diphenolic acid, succinic acid and 1,4 butanediol which are intermediates compounds employed in the manufacturing of polycarbonates and other plastics. Additionally, the LA has been used as solvent in polymer industry (Yan et al., 2015; Rackemann and Doherty, 2011). Fig. 7.2 illustrates some important derivates of LA and their respective applications. Finally, it is important to mention, that despite of the important applications of LA, these uses have been poor developed at industrial scale. The reasons are associated with low yields, high equipment and operational costs, difficult handling, and recovery during the LA production (Mukherjee et al., 2015).

7.3 Current levulinic acid markets

As previously mentioned, LA has been poor applied at an industrial level, as consequence, the current markets and demands for LA are limited in contrast to other biochemicals. The demand for LA in 2013 was estimated in 2606 tons and it is expected that for the year 2020 this the demand reached 3820 tons, which represents an annual increase of 3.5% (GVR, 2015). This growth is caused by an increasing use of LA for producing

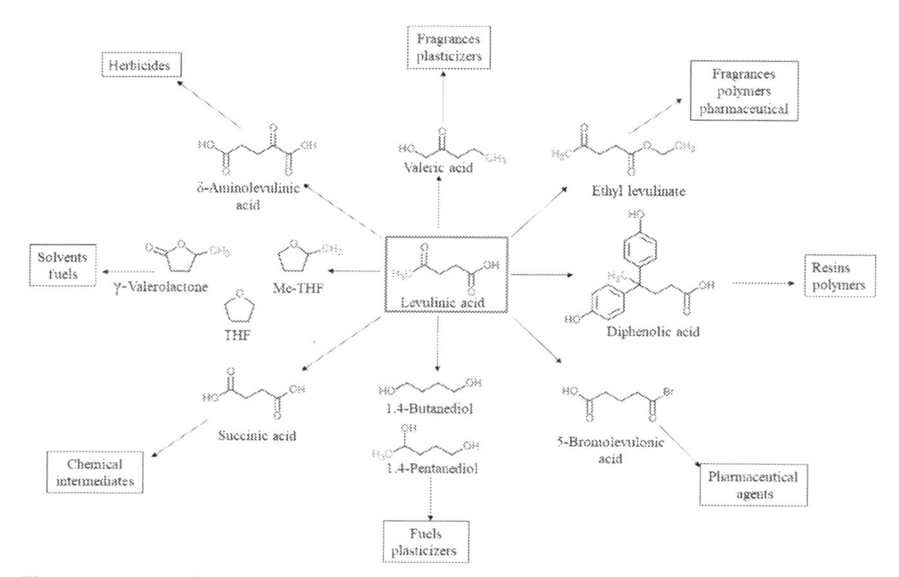

Figure 7.2 Main levulinic acid derivatives and their applications.

pesticides, pharmaceuticals, cosmetics, solvents, among other applications (GVR, 2015). Despite of limited demand for LA, some sources have estimated that the LA market is valued around USD 19.65–27.76 million in 2020 resulting in a selling price of 7267 USD/ton, which becomes it in one of the most valuable biochemicals (Taylor et al., 2015; GVR, 2015). On the other hand, it is estimated that levulinic market will have a value around USD 61.2–71.85 million by 2027, this growth will be reached at an annual growth rate of 14.10% (GNW, 2020; DBMR, 2020).

Currently, North America is considered the largest LA market due to the large demand, required by two main industries, the pharmaceutical and personal care industries. Both sectors have undergone significant growths in recent years, principally in in the United States and Canada (GVR, 2015; GNW, 2020; DBMR, 2020). On the other hand, the Asia Pacific zone is considered the second largest market of LA. Due to agrochemicals industry, also it is expected that this zone will experience the fastest growing during the following years (Mordor Intelligence, 2019). Fig. 7.3 shows the growth expectation of LA consumption from 2019 to 2024.

Today, the LA market is dominated and supplied by few companies, some of the key LA suppliers include GF Biochemicals, Avantium NV, Hefei TNJ Chemical Industry Co. Ltd., Heroy Chemical Industry Co. Ltd., and Haihang Industry Co. Ltd., DuPont, Hebei Shijiazhuang

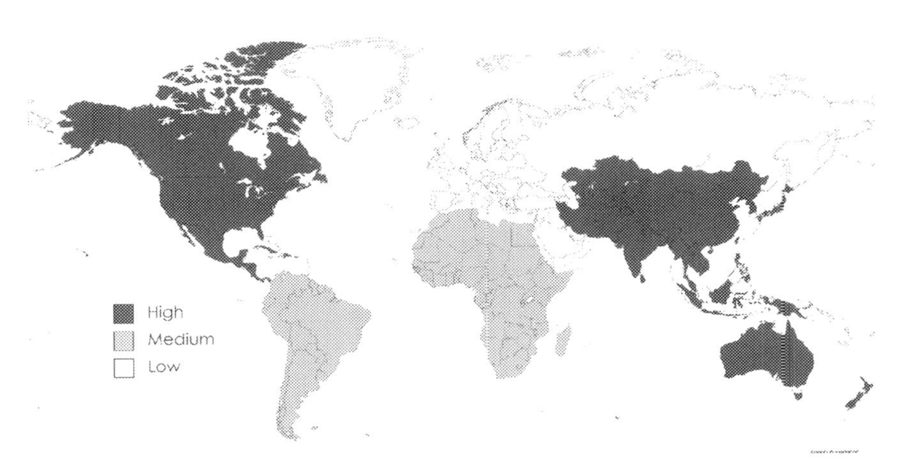

Figure 7.3 Market growth rate of levulinic acid demand by region from 2019 to 2014. *Data from Mordor Intelligence, 2019, https://www.mordorintelligence.com/industry-reports/levulinic-acid-market.*

Worldwide Furfural & Furfuryl Alcohol Furan Resin Co. Ltd., Shijiazhuang Pharmaceutical Group Ouyi Pharmaceutical Co. Ltd., Biofine Technology LLC, Segetis, Shanghai Apple Flavor & Fragrance Co. Ltd, among others. These LA plants are located mainly in the United States, China, Italy, India, Japan, and other countries, however, their total production capacity of LA is unknown (GVR, 2015; DBMR, 2020). GF Biochemicals is the major global supplier of LA, at the same time, this company has the biggest LA plant with an estimated total capacity of 10,000 tons per year located at Caserte Italy, despite of the enormous capacity of this facility, it produced only 1200 tons/year in 2015, but the firm was expected to produce 4200 tons/year for 2016, however, the company has not provided information about if this production level was reached. Finally, GF Biochemicals is wanting to build another LA facility with a capacity of 50,000 tons/year (Scott, 2016; Mordor Intelligence, 2019).

7.4 Kinetics models for levulinic acid production

As aforementioned, LA can be produced from hemicellulose or cellulose fractions of biomass. In this sense, the most common method to produce LA is through hydrolysis of biomass using dilute solutions of mineral acids at high temperature conditions (Chang et al., 2009; Chin et al., 2014). During the acid hydrolysis, the different fractions of biomass are decomposed into different products, the cellulose decomposes to LA, while the hemicellulose produced furfural. Therefore, if it is desired to produce LA from the hemicellulose fraction, an extra step of hydrogenation from furfural to LA is necessary (Adeleye et al., 2019). Therefore, the cellulose route is more preferable than the hemicellulose route, also the cellulose fraction is more abundant in the biomass, thus higher yields can be obtained by this pathway (Bozell et al., 2000). In this section, the kinetic models to produce LA by cellulose route using acid hydrolysis are commented.

Although LA is produced in a very similar way that furfural, the kinetic models and chemical mechanisms of LA formation have not well established in contrast to the furfural case. For this reason, several kinetic models with different reaction mechanisms can be found in the literature. Nevertheless, due to the production of LA is sensitive to temperature and acid concentration, the specific reaction rates (k_n) can be calculated using the modified Arrhenius expression as follows.

$$k_n = aC^m \exp\left(-\frac{E}{RT}\right) \tag{7.1}$$

Where a and m are regression parameters and C is the acid or catalyst concentration usually expressed in weight percentage (wt.%/wt.%) and E and T are the activation energy and temperature, respectively. This modified Arrhenius equation is commonly used in the LA kinetics and it has been applied successfully in numerous previous works (Girisuta et al., 2013; Chang et al., 2009; Dussan et al., 2013; Zheng et al., 2017; Chin et al., 2014). Concerning the chemical mechanism and kinetic models proposed to LA production, these models depend strongly on biomass utilized as raw material. Table 7.2 illustrates some pseudo-first order kinetic models reported in the literature. It is important to highlight that all the kinetic models presented in Table 7.2 use the modified Arrhenius equation to calculate the specific reaction rates.

7.5 Current for levulinic acid production

The production of LA from lignocellulosic biomass is not new, the first report of LA production dates since 1840 by G.J. Mulder (Morone et al., 2015). Nevertheless, the first commercial production of LA began in 1940 in the United States by A.E. Stanley Mfg. using an autoclave. This first commercial process used hydrochloric acid and starch as feedstock (Moyer, 1942). Subsequently, in 1953 the pioneering Quaker Oats Company developed and patented the first continuous process for LA production. This process was invented in order to recycle the cellulosic residues obtained during the furfural production process. However, this process had the capability of working with different raw materials such as corncobs, bagasse, and oat hull. The continuous Quaker Oats process is based on the acid hydrolysis of biomass or residue using nonvolatile mineral acids such as sulfuric acid, phosphoric acid, or sodium sulfate as catalyst, at medium-high temperatures between 150°C–200°C in order to degrade the raw materials into LA. Once the LA is produced, the excess of water is removed using a liquid-liquid extraction with methyl isobutyl ketone or furfural as solvent. Finally, the remaining impurities are removed from LA by distillation (Dunlop and Riverside Wells, 1957). A simplified process flowsheet of LA process proposed by Quaker Oats is presented in Fig. 7.4.

Despite the continuous process proposed by Quaker Oats Company was development since 1953, this process and other alternatives have not

Table 7.2 Kinetic models and parameters for levulinic acid production.

Raw material/catalyst	Proposed model	Reaction conditions	Activation energies (*kJ/moL*)	References
Corn stall/$FeCl_4$	$\text{Cellulose} \xrightarrow{k_1} \text{Glucose}_{k_2\downarrow} \text{Humins} \xrightarrow{k_3} \text{HMF} \xrightarrow{K_4} \text{LA} + \text{Formic Acid}$	$T = 160°C–200°C$ $C = 0.1–0–5\ moL/L$	$E_1 = 77.55$ $E_2 = 93.25$ $E_3 = 94.04$ $E_4 = 46.86$	Zheng et al. (2017)
Miscanthus × giganteus/H_2SO_4	$\text{Cellulose} \xrightarrow{k_1} \text{Glucose}_{k_2\downarrow} \text{Humins} \xrightarrow{k_3} \text{HMF} \xrightarrow{K_4} \text{LA} + \text{Formic Acid}$	$T = 150°C–200°C$ $C = 0.1–0–53\ moL/L$	$E_1 = 188.9$ $E_2 = 186.2$ $E_3 = 155.5$ $E_4 = 121.3$	Dussan et al. (2013)
Wheat straw/H_2SO_4	$\text{Cellulose} \xrightarrow{k_1} \text{Glucose}_{k_2\downarrow} \text{Humins} \xrightarrow{k_3} \text{HMF}_{k_5\downarrow} \text{Humins} \xrightarrow{K_4} \text{LA} + \text{Formic Acid}$	$T = 190°C–230°C$ $C = 1–5\ wt.\%$ $Biomass\ load = 5.9\ wt.\%$	$E_1 = 78.66$ $E_3 = 54.51$ $E_4 = 56.51$	Chang et al. (2009)
Corncobs/H_2SO_4	$\text{Cellulose} \xrightarrow{k_1} \text{Glucose} \xrightarrow{k_2} \text{HMF}_{k_4\downarrow} \text{Humins} \xrightarrow{K_3} \text{LA} + \text{Formic Acid}$	$T = 150°C–180°C$ $C = 0.2–0.8\ moL/L$	$E_1 = 135.32$ $E_2 = 134.36$ $E_3 = 96.51$ $E_4 = 101.58$	Liang et al. (2017)
Sugarcane bagasse/H_2SO_4	$\text{Cellulose} \xrightarrow{k_1} \text{Glucose}_{k_2\downarrow} \text{Humins} \xrightarrow{k_3} \text{HMF} \xrightarrow{K_4} \text{LA} + \text{Formic Acid}$	$T = 150°C–200°C$ $C = 1–5\ wt.\%$ $Biomass\ load = 10\ wt.\%$	$E_1 = 144.8$ $E_2 = 161.4$ $E_3 = 152.1$ $E_4 = 101.6$	Girisuta et al. (2013)
Water hyacinth/H_2SO_4	$\text{Cellulose}_{k_4\downarrow} \text{Decomposition Products} \xrightarrow{k_1} \text{Glucose}_{k_5\downarrow} \text{Humins} \xrightarrow{k_2} \text{HMF}\ \text{Humins} \xrightarrow{K_3} \text{LA} + \text{Formic Acid}\ {}_{k_6\downarrow}$	$T = 150°C–175°C$ $C = 0.1–1\ moL/L$ $Biomass\ load = 10\ wt.\%$	$E_1 = 151.5$ $E_2 = 152.2$ $E_3 = 110.5$ $E_4 = 11.3$ $E_5 = 164.7$	Girisuta et al. (2008)

Sugarcane bagasse/H_2SO_4	Cellulose $\xrightarrow{k_1}$ Glucose$_{k_2\downarrow}$ Humins $\xrightarrow{k_3}$ HMF $\xrightarrow{K_4}$ LA + Formic Acid	$T = 150°C–190°C$ $C = 3–7\ wt.\%/v$	$E_1 = 57.50$ $E_2 = 32.64$ $E_3 = 37.60$ $E_4 = 24.42$	Lopes et al. (2020)
Rice husk/H_2SO_4	Cellulose $\xrightarrow{k_1}$ Glucose$_{k_2\downarrow}$ Humins $\xrightarrow{k_3}$ HMF $\xrightarrow{K_4}$ LA + Formic Acid	$T = 150°C–190°C$ $C = 3–7\ wt.\%/v$	$E_1 = 57.50$ $E_2 = 3.62$ $E_3 = 6.1$ $E_4 = 24.42$	Lopes et al. (2020)
Soybean straw/H_2SO_4	Cellulose $\xrightarrow{k_1}$ Glucose$_{k_2\downarrow}$ Humins $\xrightarrow{k_3}$ HMF $\xrightarrow{K_4}$ LA + Formic Acid	$T = 150°C–190°C$ $C = 3–7\ wt.\%/v$	$E_1 = 57.50$ $E_2 = 4.14$ $E_3 = 6.03$ $E_4 = 24.42$	Lopes et al. (2020)

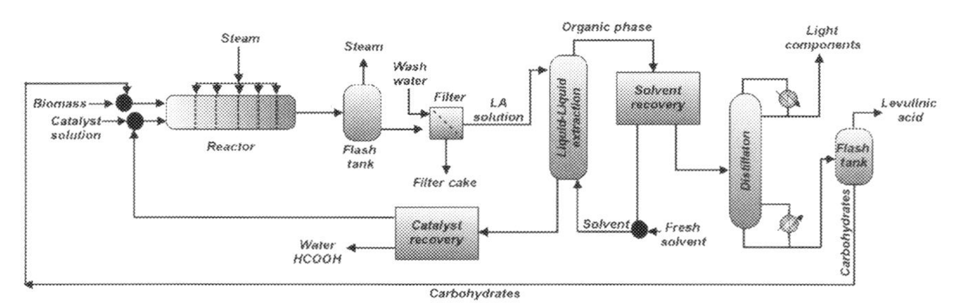

Figure 7.4 Continuous process for levulinic acid production proposed by Quaker Oats Company (Dunlop and Riverside Wells, 1957).

reached a significant industrial and commercial scale, the motives are related to expensive operational costs and low yields (Mukherjee et al., 2015). Considering those challenges into account, the efforts have focusing on the development of a more efficient and profitable process for LA production. As a result of those efforts in 1999, the biofine process was invented by Stephen W. Fitzpatrick (Fitzpatrick, 1990). Nowadays, the biofine process is considered as the most efficient and profitable technology to produce LA from lignocellulosic biomass (Morone et al., 2015). Similarly, to Quaker Oats' continuous process, the biofine technology is based on the rapid hydrolysis of biomass using dilute solutions of mineral acids. However, some important differences during the acid hydrolysis of biomass become the biofine process in a more efficient alternative. The technology differs from other processes owing to it performs the biomass hydrolysis in two reactors, which allows overcoming the limitations in LA production (Hayes et al., 2006; Mukherjee et al., 2015). This first stage in a biofine plant is the mixing of biomass with a dilute solution of sulfuric acid, the concentration of this solution should be around 1%–5% depending on biomass. Then the mixture is fed to a plug flow reactor, where the carbohydrates contained in the biomass are converted to soluble intermediates such as 5 hydroxymethylfurfural (HMF), this conversation is carried out at thermal condition about 210°C–220°C and 25 bar. Because the reaction hydrolysis occurs very fast, the required residence time in this reactor is only 12 s. Subsequently, the outlet stream of first reactor is sent to a second reactor, which operates with less severe conditions of 190°C–200°C and 14 bars, for this reason, the second reactor requires a resident time of 20 min. During this stage, furfural and other light compounds are removed, whereas the LA and other residues are sent to a gravity separator or a filter. Finally, the

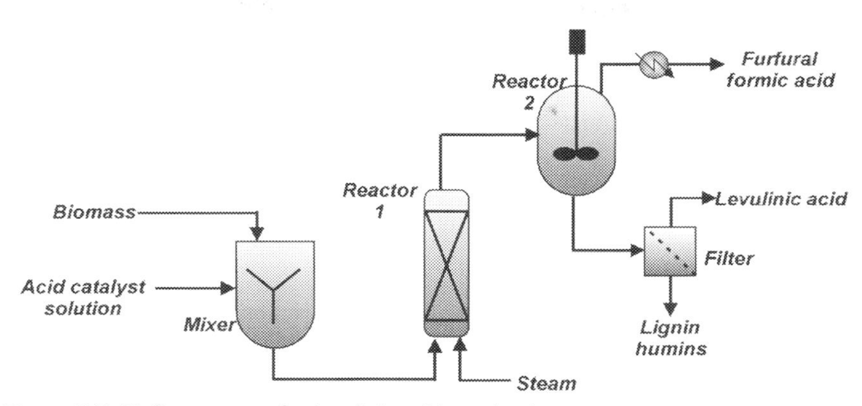

Figure 7.5 Biofine process for levulinic acid production.

impurities of LA are removed by distillation (Hayes et al., 2006). Fig. 7.5 shows a simplified diagram of the biofine process. It is important to mention, that a real plant could require another additional process stage, this will depend on the desired final products. Because of the more efficient reactor system of biofine process, it can achieve yields around 80% of the maximum theoretical yield. The improvements achieved in the biofine technology, become it in the first profitable and commercially viable LA process. As a proof of this, in 2005 the GF Biochemicals company used this technology to build the biggest LA plant at Caserte Italy, this factory has a theoretical capacity of 10,000 tons/year of LA (Scott, 2016; Hayes et al., 2006).

7.6 New intensified proposals for levulinic acid production

During several years, the efforts and advances in LA production have been focused in improve the hydrolysis of biomass in order to increase, both the LA yield and the process profitability. The biofine process is the result of those efforts, due to it is the first commercially viable process. This process bases its profitability on more efficient conversion of biomass into levulinic. However, few works are focused on improving the separation stage in LA production. The purification stage is one of the most critical and expensive stages in a biorefinery, due to the large amounts of water that must be removed. In this sense, the biofine process is not the exception, since it requires significant quantities of water to achieve suitable hydrolysis of biomass. As a consequence, large amounts of water must be separated from LA; thus, an intensive use of energy is required.

Therefore, the profitability of LA production can be improved by a reduction of energy consumption on the separation stage. In this sense, the process intensification seems to be a suitable tool to reduce the energy consumption in LA purification.

The first work that addressed the LA purification was presented by Dunlop (1954), he suggested the use of liquid-liquid extraction using immiscible solvents to remove the massive amount of water. Additionally, Dunlop (1954) proposed the methyl isobutyl ketone, furfural, methyl chloride and other halogenated hydrocarbons as suitable solvents that may be used to purify LA. Subsequently, Dunlop and Riverside Wells (1957) patented a separation process which uses methyl isobutyl ketone as entrainer. Afterward, Seibert (2010) invented an extractive process to purify LA and formic acid contained in the aqueous mixture from an acid hydrolysis stage. The process proposed by Seibert (2010) consists of a liquid-liquid extraction following by three distillation columns. The main advantage of this process is the use of furfural as a solvent which is one of the main products of biomass' acid hydrolysis, reducing in this way the cost of the process. On the other hand, the first intensified proposed in the LA purification was presented by Nhien et al. (2016), their proposal is based on the process proposed by Seibert (2010), however, they replaced two distillation columns with a dividing wall column in order to reduce the energy requirements. Additionally, Nhien et al. (2016) studied other solvents to purify LA, but they found that furfural is the most suitable option. Fig. 7.6 shows a comparative of the purification processes proposed by Dunlop and Riverside Wells (1957), Seibert (2010) and Nhien et al. (2016).

Recently, Alcocer-García et al. (2019) reported one of the most important works about process intensification and optimization for the purification of LA. Their work consists of the synthesis and optimization of several intensified process alternatives. All the intensified options were generated from the previous process proposed by Seibert (2010), which was used as a benchmark to measure the improvements providing by the intensified alternatives. Alcocer-García et al. (2019) considered a feed stream of 90,000 kg/h with a mass composition of 86% water, 7% LA, 4% furfural and 3% formic acid at 298.15 K and 202.95 kPa. These compositions o correspond to compositions produced by typical acid hydrolysis reactors (Nhien et al., 2016). The mixture was modeled using the Non-random Two-Liquids model coupled to Hayden-O'Connell equation of state (NRTL-HOC), in order to consider the dimerization and solvation

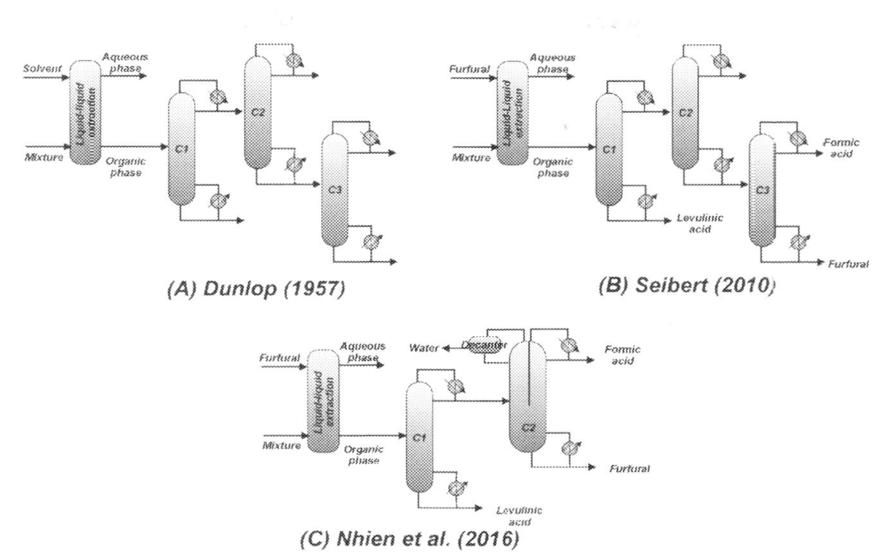

Figure 7.6 Separation schemes for levulinic acid purification.

of mixtures with carboxylic acids compounds. All the intensified processes were designed considering the following purities: 98% (wt.%) for LA, 85% for formic acid and 99.9% for furfural those purities correspond to the minimum purity specifications required by current industrial applications.

Four different separation options were studied by Alcocer-García et al. (2019), these configurations consist of a conventional scheme (CS) taking as reference the hybrid extractive distillation process proposed by Seibert (2010). Then the intensified dividing wall column scheme with a decanter (DWCS-D) proposed by Nhien et al. (2016) is studied. Finally, two novel intensified options are proposed, a dividing wall column configuration with a decanter and thermal couplings (DWCS-DA) and a double dividing wall column scheme with a decanter (TDWS-D). The synthesis of the intensified diving columns was performed according to the procedure of thermodynamic equivalent configurations proposed by Errico et al. (2009). The methodology is quite simple, and it consists of four main steps, which are described below:

1. Identity a possible conventional distillation sequence, in this case, the process proposed by Seibert (2010) is the conventional one. Then the column section must be classified. A column section is described as a part of a column that is not interrupted by inlet or outlet mass streams

or heat flows. With this in mind, the conventional distillation columns can be divided into two or more different sections.

2. Once the column sections have been identified in the CS, a thermally coupled configuration is generated by substituting one or more condenser and reboilers related to a not product stream with interconnecting liquid a vapor stream.
3. Using the thermally coupled schemes as a reference, the thermodynamically equivalent alternatives are derived through moving a column section related to a condenser or a reboiler, that supplies a common reflux flowrate or the vapor boil up among two successive columns. Subsequently, the procedure continues with the elimination of a condenser or a reboiler correlated with a submixture from a column in a distillation sequence, which becomes the rectifying or stripping sections of the successive column movable.
4. Finally, two or more stripping sections of an intensified scheme can be incorporated into a singles shell. This procedure can be used to generate different intensified separation alternatives. Fig. 7.7 summarizes the procedure applied to generate the DWCS-D scheme, the other intensified options were designed in similar way. The complete alternatives studied by Alcocer-García et al. (2019) are shown in Fig. 7.8.

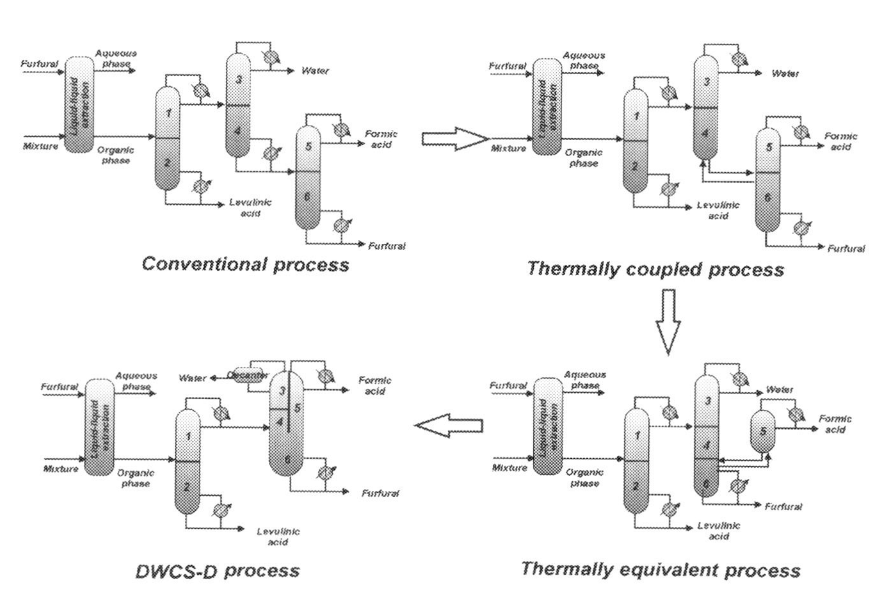

Figure 7.7 Synthesis procedure for the intensified dividing wall processes.

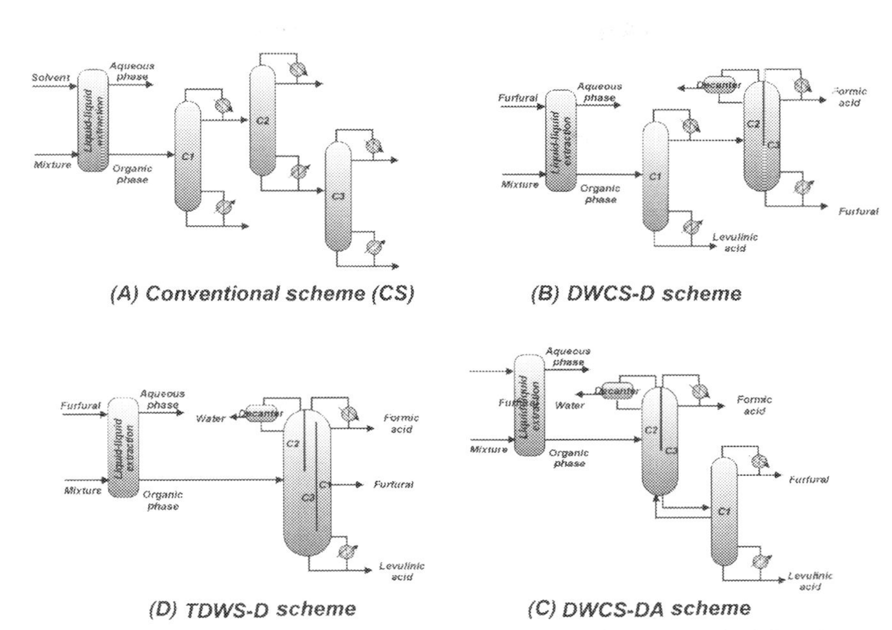

Figure 7.8 Distillation sequences for levulinic acid purification studied by Alcocer-García et al. (2019).

The intensified processes were simultaneously designed and optimized using the Differential Evolution with Tabu List (DETL) technique proposed by (Srinivas and Rangaiah, 2007). The optimization technique was executed using a hybrid platform that links Excel through Visual Basic with Aspen Plus. In this hybrid platform, the optimization method is programmed and executed in Visual Basic within Excel, while Aspen Plus performs the rigorous calculations of solutions proposed by the DETL method. The Total Annual Cost (TAC) and Eco-Indicator 99 were considered as metrics to evaluate the economic and environmental performance of solutions during the optimization procedure. These objective functions were explained in depth during the last chapter, for this reason in this chapter only the objective function is explained.

Mathematically, the multi- objective optimization problem is expressed as follows:

$$\begin{gathered} \min[TAC, EI99] = f(S, N_c, N_i, NF_i, RR_i, D, FL, FV_i, DC_i) \\ \text{Subject to:} \\ \vec{y}_m \geq \vec{x}_m \\ \vec{w}_m \geq \vec{y}_m \end{gathered} \tag{7.2}$$

where is S is the mas flowrate of solvent (kg/hour), N_e is the number of stages of an extractive column, N_i is the number of stages of column i, NF_i is the feed stage for the distillation column i, RRi is the reflux ratio of column i, D is the distillate flowrate (kg/hour), FL is the interconnecting liquid flowrate (kg/hour), FV is the interconnecting vapor flowrate (kg/hour), DC_i the diameter of column i. The optimization was carried out in order to satisfy the purity and mass flowrate constraints for the different compounds. The parameters used for the DETL method are number of population (NP): 120 individuals, Generations Number (GenMax): 417, Tabu List size: 60 individuals, Tabu Radius: 0.00001, Crossover fractions (Cr): 0.8, Mutation factor (F): 0.6. The decision variables considered for each distillation scheme are presented in Table 7.3.

The results obtained by Alcocer-García et al. (2019) are shown in the Pareto fronts charts of Fig. 7.9. The results presented were obtained after 50,000 iterations. It is important to highlight, that those iterations were simulated by the optimization algorithm using the RADFRAC block contained in Aspen Plus. Therefore, it can be considered that the processes were robustly designed, due to this block contains the complete set of mass-energy balances and equilibrium equations. The utopian point methodology was used to select the solution with the best trade-off between the objectives for each Pareto front, this solution is representing as a big dot. According to the information provided by Alcocer-García

Table 7.3 Decision variables using during the optimization procedure.

Decision variables	CS	DWCS-D	DWCS-DA	TDWS-D
Number of stages extractive column	X	X	X	X
Number of stages C1, C2, C3	X	X	X	X
Feed stage C1, C2, C3	X	X	X	X
Reflux ratio C1	X	X	–	–
Reflux ratio C2	X	–	–	–
Reflux ratio C3	X	X	X	X
Distillate rate. C1	X	X	X	X
Distillate rate. C2	X	X	X	
Distillate rate. C3	X	X	X	X
Diameters C1, C2, C3	X	X	X	X
Interconnecting steam C1	–	–	X	X
Interconnecting steam C4	–	X	X	X
Interconnecting steam C4				X
Solvent mass flowrate	X	X	X	X

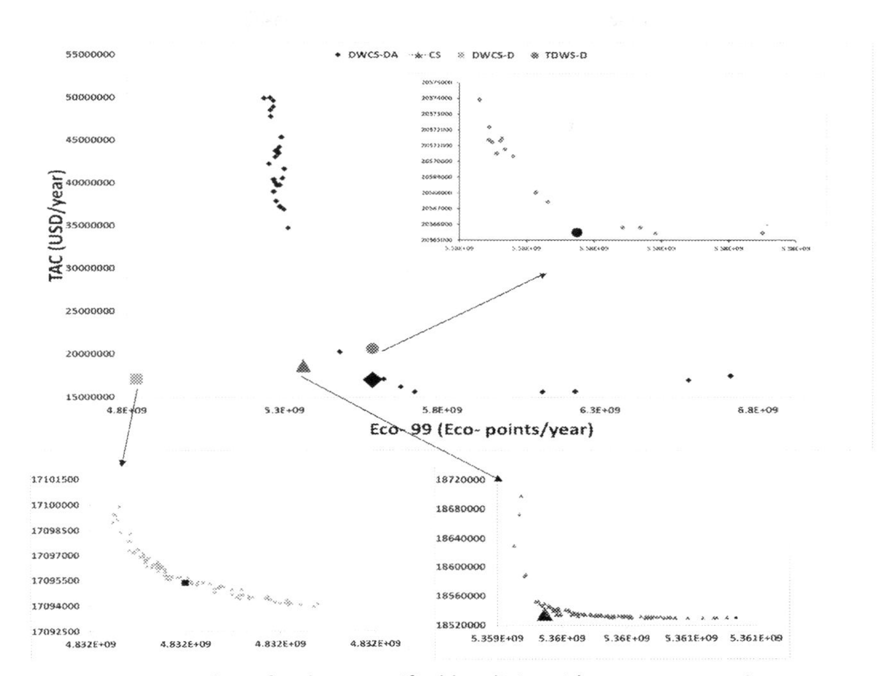

Figure 7.9 Pareto front for the intensified levulinic acid process separation.

et al. (2019), all the designs have similar values in some design specifications, such as the number of stages or reflux ratios, hence the differences in TAC are caused by the diameter of the columns. As a consequence, when the diameter is increased the TAC increases too. In relation to EI99, this parameter is influenced by the energy requirements. Please note, that some intensified designs do not provide TAC and EI99 savings in contrast to some conventional designs, the reason is related to the internal flows required by the dividing wall column. In the case of TDWS-D and DWCS-DA, the processes topology favors an increment on internal flows and thus an increment on energy consumption. Once the large amounts of water have been removed during extraction, the LA becomes in the most abundant compound in the mixture. For this reason, it is thermodynamically more efficient to remove the LA first and then the other coproducts. The priority remotion of LA occurs in both conventional and DWCD-D alternatives, thus the energy consumption and environmental impact are less in these sequences.

The results indicate that the DWCS-D process is the best intensified alternative to purify LA. This alternative presents the highest environmental improvements in contrast to the other two intensified alternatives, also it has TAC similar TAC values to some conventional designs. In this case, owing to the mixture has huge amounts of water, an important quantity of energy is spent in removing the water. Therefore, the energy savings provided by the intensified processes are not significant, in contrast, some intensified processes are more expensive than some conventional designs (i.e., TDWS-D process). Those poor costs savings have been observed in other intensified separations of products derived from acid hydrolysis, such as furfural, hence the efforts and improvements of LA process must be focused on a cost-effective water remotion.

7.7 Conclusions

This chapter addressed some important features, data, and advances about LA production. Although the demand for LA is not so high as other biochemicals, it is expected that in the following years, LA will become in one of the most important biochemicals. For this reason, important advances in LA production are required, in order to reduce the production costs and increase its competitiveness concerning other current chemicals. Today, the biofine process is the only profitable alternative to produce LA, therefore the research efforts should be focused on the development of new processes or improving the process purification of LA. In this sense, the pioneering works reported by Alcocer-García et al. (2019) and Nhien et al. (2016) contain valuable information about future advances in LA purification. Based on their results, it is required the development and testing of other technologies, in order to achieve a cost-effective way of reducing the water from the mixture and improve the profitability and environmental impact of the process.

References

Adeleye, A.T., Louis, H., Akakuru, O.U., Joseph, I., Enudi, O.C., Michael, D.P., 2019. A Review on the conversion of levulinic acid and its esters to various useful chemicals. Aims Energy 7 (2), 165–185.

Alcocer-García, H., Segovia-Hernández, J.G., Prado-Rubio, O.A., Sánchez-Ramírez, E., Quiroz-Ramírez, J.J., 2019. Multi-objective optimization of intensified processes for the purification of levulinic acid involving economic and environmental objectives. Chem. Eng. Process 136, 123–137.

Bozell, J.J., Moens, L., Elliott, D.C., Wang, Y., Neuenscwander, G.G., Fitzpatrick, S.W., et al., 2000. Production of levulinic acid and use as a platform chemical for derived products. Resour. Conserv. Recycl. 28 (3–4), 227–239.

Chang, C., Xiaojian, M.A., Peilin, C.E.N., 2009. Kinetic studies on wheat straw hydrolysis to levulinic acid. Chin. J. Chem. Eng. 17 (5), 835–839.

ChemicalBook, 2021. https://www.chemicalbook.com/ChemicalProductProperty_EN_CB3213533.htm (Revised January 2021).

Chin, S.X., Chia, C.H., Fang, Z., Zakaria, S., Li, X.K., Zhang, F., 2014. A kinetic study on acid hydrolysis of oil palm empty fruit bunch fibers using a microwave reactor system. Energy Fuels 28 (4), 2589–2597.

Data Bridge Market Research (DBMR), 2020. https://www.databridgemarketresearch.com/reports/global-levulinic-acid-market

Dunlop, A.P., 1954. *Recovery of levulinic acid*. United States Patent 2684982.

Dunlop, A.P., Riverside Wells, P.A., 1957. *Process for producing levulinic acid*. United States Patent 2813900.

Dussan, K., Girisuta, B., Haverty, D., Leahy, J.J., Hayes, M.H.B., 2013. Kinetics of levulinic acid and furfural production from Miscanthus × giganteus. Bioresour. Technol. 149, 216–224.

Errico, M., Rong, B.G., Tola, G., Turunen, I., 2009. A method for systematic synthesis of multicomponent distillation systems with less than N-1 columns. Chem. Eng. Process 48 (4), 907–920.

Fang, Z., Qi, X. (Eds.), 2017. Production of Platform Chemicals From Sustainable Resources. Springer, Singapore.

Fitzpatrick, S.W., 1990. *Production of levulinic acid from carbohydrate-containing materials*. United States Patent 5,608,105.

Galletti, A.M.R., Antonetti, C., De Luise, V., Licursi, D., Nassi, N., 2012. Levulinic acid production from waste biomass. BioResources 7 (2), 1824–1835.

Girisuta, B., 2007. *Levulinic acid from lignocellulosic biomass*. University Library Groningen [Host].

Girisuta, B., Danon, B., Manurung, R., Janssen, L.P.B.M., Heeres, H.J., 2008. Experimental and kinetic modelling studies on the acid-catalysed hydrolysis of the water hyacinth plant to levulinic acid. Bioresour. Technol. 99 (17), 8367–8375.

Girisuta, B., Dussan, K., Haverty, D., Leahy, J.J., Hayes, M.H.B., 2013. A kinetic study of acid catalysed hydrolysis of sugar cane bagasse to levulinic acid. Chem. Eng. J. 217, 61–70.

Globe News Wire (GNW), 2020. https://www.globenewswire.com/news-release/2020/08/31/2086279/0/en/Levulinic-Acid-Market-To-Reach-USD-61-2-Million-By-2027-Reports-and-Data.html.

Grand View Research (GVR), 2015. https://www.grandviewresearch.com/press-release/global-levulinic-acid-market.

Hayes, D.J., Fitzpatrick, S., Hayes, M.H., Ross, J.R., 2006. The biofine process—production of levulinic acid, furfural, and formic acid from lignocellulosic feedstocks. In: Kamm, B., Gruber, P.R., Kamm, M. (Eds.), Biorefineries-Industrial Processes and Products: Status Quo and Future Directions. WILEY-VCH Verlag GmbH & Co. KGaA, pp. 139–164. Available from: https://doi.org/10.1002/9783527619849.ch7.

Kang, S., Fu, J., Zhang, G., 2018. From lignocellulosic biomass to levulinic acid: a review on acid-catalyzed hydrolysis. Renew. Sustain. Energy Rev. 94, 340–362.

Liang, C., Hu, Y., Guo, L., Wu, L., Zhang, W., 2017. Kinetic study of acid hydrolysis of corncobs to levulinic acid. BioResources 12 (2), 4049–4061.

Lopes, E.S., Rivera, E.C., de Jesus Gariboti, J.C., Feistel, L.H.Z., Dutra, J.V., Maciel Filho, R., et al., 2020. Kinetic insights into the lignocellulosic biomass-based levulinic acid production by a mechanistic model. Cellulose 27, 5641–5663.

Mordor Intelligence, 2019. https://www.mordorintelligence.com/industry-reports/levulinic-acid-market

Morone, A., Apte, M., Pandey, R.A., 2015. Levulinic acid production from renewable waste resources: bottlenecks, potential remedies, advancements and applications. Renew. Sustain. Energy Rev. 51, 548–565.

Moyer, W.W., 1942. *Preparation of levulinic acid.* United States Patent 2270328.

Mukherjee, A., Dumont, M.J., Raghavan, V., 2015. Sustainable production of hydroxymethylfurfural and levulinic acid: challenges and opportunities. Biomass Bioenergy 72, 143–183.

National Renewable Energy Laboratory (NREL), 2004. https://www.nrel.gov/docs/fy04osti/35523.pdf

Nhien, L.C., Long, N.V.D., Kim, S., Lee, M., 2016. Design and assessment of hybrid purification processes through a systematic solvent screening for the production of levulinic acid from lignocellulosic biomass. Ind. Eng. Chem. Res. 55 (18), 5180–5189.

Pileidis, F.D., Titirici, M.M., 2016. Levulinic acid biorefineries: new challenges for efficient utilization of biomass. ChemSusChem 9 (6), 562–582.

Rackemann, D.W., Doherty, W.O., 2011. The conversion of lignocellulosics to levulinic acid. Biofuels Bioprod. Bioref. 5 (2), 198–214.

Scott, A., 2016. Opening the door to levulinic acid. C&EN 94 (2), 18–19.

Seibert, F., 2010. *A method of recovering levulinic acid.* WO Patent App. PCT.

Signoretto, M., Taghavi, S., Ghedini, E., Menegazzo, F., 2019. Catalytic production of levulinic acid (LA) from actual biomass. Molecules 24 (15), 2760.

Srinivas, Rangaiah, 2007. Industrial & Engineering Chemistry Research 46, 7126–7135. Available from: https://doi.org/10.1021/ie070007q.

Taylor, R., Nattrass, L., Alberts, G., Robson, P., Chudziak, C., Bauen, A., et al., 2015. *From the sugar platform to biofuels and biochemicals: final report for the European Commission Directorate-General Energy.* E4tech/Re-CORD/Wageningen UR.United States2009/056296.

Van de Vyver, S., Geboers, J., Jacobs, P.A., Sels, B.F., 2011. Recent advances in the catalytic conversion of cellulose. ChemCatChem 3 (1), 82–94.

Werpy, T., & Petersen, G., 2004. *Top value added chemicals from biomass: volume I—results of screening for potential candidates from sugars and synthesis gas (No. DOE/GO-102004–1992).* National Renewable Energy Lab., Golden, CO.

Yan, K., Jarvis, C., Gu, J., Yan, Y., 2015. Production and catalytic transformation of levulinic acid: a platform for speciality chemicals and fuels. Renew. Sustain. Energy Rev. 51, 986–997.

Zheng, X., Zhi, Z., Gu, X., Li, X., Zhang, R., Lu, X., 2017. Kinetic study of levulinic acid production from corn stalk at mild temperature using $FeCl_3$ as catalyst. Fuel 187, 261–267.

CHAPTER 8

Ethyl levulinate

Contents

8.1 Introduction

Ethyl levulinate (EL) is an organic compound with the chemical formula $C_7H_{12}O_3$, it is considered by the US National Renewable Energy Laboratory (NREL) as one of the 12 most promising biochemical platforms derived from biomass (NREL, 2004). On the other hand, EL together with 2-methyltetrahydrofuran are considered as two of the most important derivates of levulinic acid (LA), owing to their several applications, but principally by their use as fuel additives (Leal Silva et al., 2018). Some previous works have determined how the use of EL as fuel additive could drastically reduce, the environmental impact and pollution (Mukherjee et al., 2015). EL is a levulinate ester which is produced by the esterification reaction of LA and alcohol, using mainly a mineral acid as a catalyst. Usually, sulfuric acid or chloride acid are the most common catalysts used in the synthesis of EL, and other levulinate esters from LA (Corma et al., 2007). However, because of the homogeneous acid catalysts are difficult to separate and reuse, in recent years, other novel heterogeneous acid catalysts have been tested to produce EL, obtaining yields up to 95% (mol basis) (Fang and Qi, 2017). Fig. 8.1 illustrates the general reaction of levulinate esters using acid catalysts.

$$\text{LA} + ROH \underset{}{\overset{H^+}{\rightleftharpoons}} \text{levulinate ester} + H_2O$$

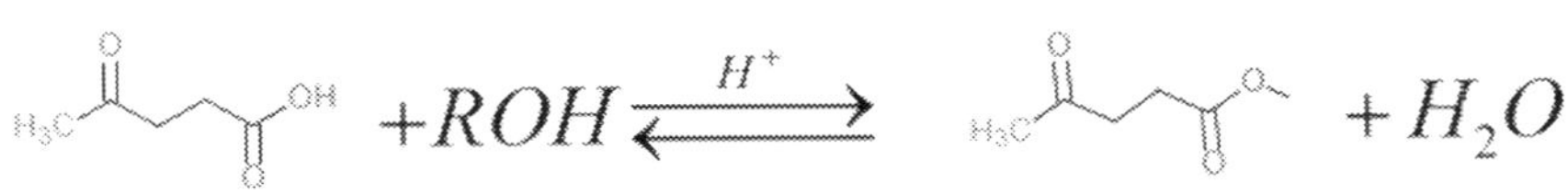

Figure 8.1 Production of levulinate esters using acid catalysts.

Improvements in Bio-Based Building Blocks Production Through Process Intensification and Sustainability Concepts
DOI: https://doi.org/10.1016/B978-0-323-89870-6.00010-X

It is important to highlight, that the chemical reaction showed in Fig. 8.1 is not the only route to produce EL and other esters. They can also be produced directly from the biomass sugars (biomass monosaccharides) such as glucose and xylose. This route requires the addition of an alcohol during the acid hydrolysis of biomass. However, this pathway is not preferable, due to some present drawbacks, like large energy demand and the formation of nondesired subproducts such as humins, which reduces significantly the EL yields (Ahmad et al., 2016; Fang and Qi, 2017).

Similarly, to other biochemicals, the production and synthesis of EL is not new, it has been synthesized since XIX. However, up to now a feasible and profitable large-scale process has not developed. The reason is associated with the production of LA. EL is a derivative of LA, as a consequence, its successful industrial production depends strongly on efficient and profitable manufacture of LA (Leal Silva et al., 2018). Nevertheless, in recent years and thanks to the development and industrial implementation of the biofine process, which is the first economically feasible process to produce LA, the efforts to development an efficient and economically viable EL process have been renewed (Leal Silva et al., 2018). As proof of the aforementioned, new pathways and technologies have emerged, such as the new superacidic zirconia catalysts or the use of ionic liquids as catalysts (Ramli and Amin, 2017; Yadav and Yadav, 2014).

8.2 Current applications and markets of ethyl levulinate

In recent years, the EL has attracted interest because of its potential to be used as a fuel additive in diesel blendings. However, more recently the EL has been applied for producing medicines as calcium levulinate, which is used for both mineral supplement and tuberculosis treatments. In addition, EL is considered as a potential substitute for valencene in the manufacturing of fragrances and fruity, sweet, and floral flavors (Leal Silva et al., 2018).

In relation to the current EL markets, since EL is a product derived from LA, its potential markets are even further limited. In this sense, Grand View Research (GVR) (2016) estimated that in 2014 the global EL market had a value of USD 8.8 million, also for the period between 2015 and 2026, a growing about 3%–3.6% is expected, which means that for 2026 the EL market will reach the USD 14 million (EMR, 2020; GVR, 2016).

Currently, EL market is at elementary and nascent status, due to its recent production technology, for this reason, its market is dominated by a

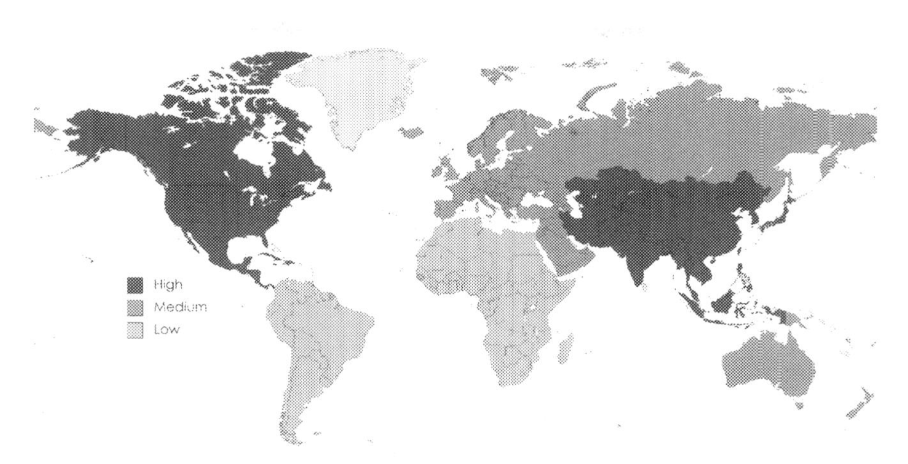

Figure 8.2 Market growth rate of ethyl levulinate demand by region from 2020 to 2025. *Mordor Intelligence (2019). (Data from Mordor Intelligence (2019) https://www.mordorintelligence.com/industry-reports/ethyl-levulinate-market.)*

few companies such as GFBiochemicals Ltd, Vigon International Inc., Ernesto Ventós, SA among others (GMI, 2020; Mordor Intelligence, 2019). Nevertheless, it is estimated that in the next few years, the demand for LA will increase and therefore the market will reach maturity. Additionally, it is expected that the production costs will decrease owing to improvements in the production process, economies of scale factors and more acceptability of this product in different industries (GVR, 2016).

Finally, it is important to highlight, that according to different sources the current LA market is dominated by the Asia-Pacific Region, which represents about 30% of the total LA market in 2014 (GVR, 2016; Mordor Intelligence, 2019). In addition, it is expected that this region will have one of the fastest growing demands of EL, due to the continuous growth of some key sectors such as the fragrance, food, and fertilizer industries, mainly in China and India (GVR, 2016; Mordor Intelligence, 2019). In this sense, Fig. 8.2 shows the market growth rate by region for the period 2020–2025. Please note that the North America region is considered as another important EL market.

8.3 Kinetics models for ethyl levulinate production

As mentioned earlier, the EL can be produced using different pathways and catalysts, consequently, there are different kinetic models to express chemical kinetics in the production of EL. However, the esterification of

LA with ethanol to produce EL stands up over other routes because this reaction can be effortlessly implemented and, at the same time, the esterification is cheaper than other pathways, despite its equilibrium limitations, which can be overcome using reactive distillation schemes (Novita et al., 2017). The reversible esterification reaction to produce EL using amberlyst as proposed by Tsai (2014) is shown in Eq. (8.1).

$$\underset{A}{C_5H_8O_3} + \underset{B}{C_2H_5OH} \underset{k_2}{\overset{k_1}{\rightleftarrows}} \underset{C}{C_7H_{12}O_3} + \underset{D}{H_2O} \tag{8.1}$$

where A = LA; B = ethanol; C = EL, and D = water. Although the esterification reaction is carried out on a solid catalyst, the load of said catalyst is so small that the reaction can be modeled as nonideal quasihomogeneous model (Tsai, 2014). Therefore, the kinetic model of reversible reaction can be expressed as an elementary reaction as follows:

$$-r_A = k_1\left(a_A a_B - \frac{a_C a_D}{K_a}\right) \tag{8.2}$$

where K_a is the equilibrium constant, which can be expressed as the ratio among the specific reaction rates or concentrations of the forward and reverse reactions:

$$K_a = \frac{k_1}{k_2} = \frac{C_C C_D}{C_A C_B} \tag{8.3}$$

The specific reaction rates k_1 and k_2 are expressed using the Arrhenius equation as follows:

$$k_1 = A_f \exp\left(\frac{-E_{0,f}}{RT}\right) \tag{8.4}$$

$$k_2 = A_r \exp\left(\frac{-E_{0,r}}{RT}\right) \tag{8.5}$$

Table 8.1 shows the kinetic parameters for EL production using amberlyst as proposed by Tsai (2014).

8.4 Current technologies for ethyl levulinate production

Today, the production of EL is dominated by a few numbers of companies, however, there is not information related to the current technology used to produce LA. Despite this lack of information one of the most

Table 8.1 Kinetic parameters for ethyl levulinate production using amberlyst.

Parameter	Value
A_f (mol/s · kg)	0.0133×10^7
A_r (mol/s · kg)	3.077×10^3
E_f (kJ/mol)	37.79
E_r (kJ/mol)	36.91

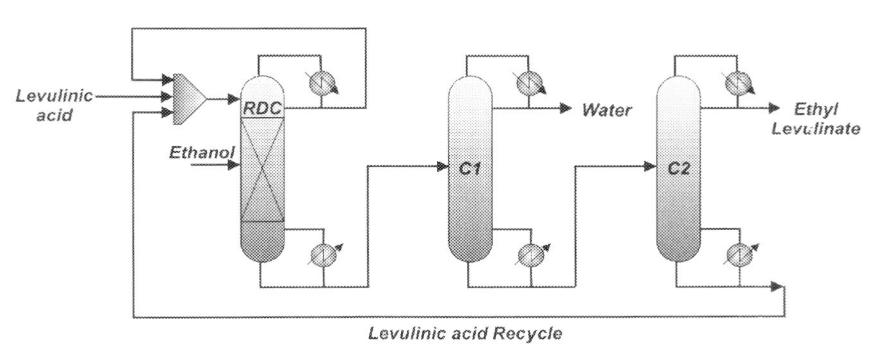

Figure 8.3 Conventional reactive distillation process proposed by Novita et al. (2017).

promising technologies to produce EL is the reactive distillation RD, where two different operations; the reaction and separation are carried out in single equipment. The use of reactive distillation in the production EL has attracted attention in recent years, because it allows to overcome the equilibrium conversion limitation. In this sense, the first record of reactive distillation applied to EL production was proposed by Novita et al. (2017). Fig. 8.3 illustrates the reactive distillation process proposed by Novita et al. (2017).

The process proposed by Novita consists of three different columns, the first column is the reactive distillation column (RDC), it has 51 stages, and the ethanol is the limiting reactive. In this column 100 kmol/hour are fed join to 126 kmol/hour of LA, which corresponds to 100 kmol/hour of fresh LA feed and 26 kmol/hour from the recirculation stream. This column has an energy requirement of 3.68 Gcal/hour. The second column (C1) is a prefractionation column, this unit aims to separate the water from the other products, it has 10 stages and an energy consumption of 2.37 gcal/hour. Finally, the third column (C2) separates the EL from the unreacted LA with a purity 99.5% mol basis. This unit has

17 equilibrium stages and energy consumption of 1.72 Gcal/hour. It is important to mention, that this process is known as Conventional Reactive Distillation Process (CRDP).

8.5 Current advances in ethyl levulinate production

As mentioned earlier, the reactive distillation is considered the most feasible process to produce EL because it allows to overcome all the equilibrium limitations associated with esterification reactions. Although reactive distillation is already an intensified process in itself, it can still be improved by increasing the intensification even further. In this sense, thermally coupled and heat integration schemes could be attractive alternatives for improving the production of EL by reactive distillation. With this in mind, Novita et al. (2017) proposed the first work about intensification to reactive distillation process for EL production. They proposed three intensified reactive distillation alternatives using as based case the conventional arrangement showed in Fig. 8.3, their intensified options consist in a thermally coupled reactive distillation (TCRD), a reactive distillation with heat integration (RDHI) and a hybrid heat integration–thermally coupling arrangement (THRD). Subsequently, Vázquez-Castillo et al. (2019) proposed an additional intensified option, which consists of a reactive distillation column followed by a dividing wall column (PDWC). The said intensified purposes are shown in Fig. 8.4.

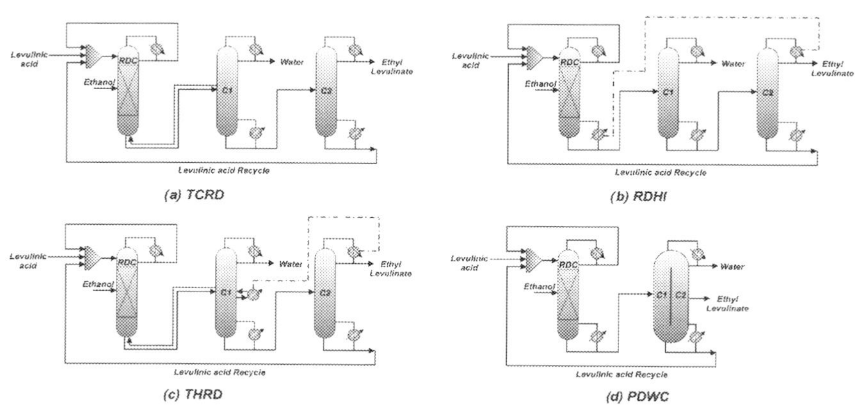

Figure 8.4 Intensified reactive distillation processes for ethyl levulinate production. a) Thermally Coupled Reactive Distillation (TCRD), b) Reactive Distillation with Heat Integration (RDHI), c) Thermally Coupled with Heat Integration Scheme (THRD) d) Divided Wall Column Scheme (PDWC).

In order to determine the optimal solution and best intensified option, Vázquez-Castillo et al. (2019) performed the simultaneous design and optimization of the aforementioned reactive distillation options. They used the differential evolution with the tabu list optimization method to design and optimize the alternatives, they considered the total annual cost (TAC), eco-indicator 99 (EI99) and individual risk (IR) as metrics to quantify the performance of different solutions. Those indices are explained in depth on Chapter 6, Furfural. Mathematically, the optimization problem proposed by Vázquez-Castillo et al. (2019) can be expressed as follows:

$$\min Z = \{TAC; IR; ECO99\} = \left\{ OPEX + \frac{CAPEX}{Payback\ period}; \sum\nolimits_b \sum\nolimits_d \sum_{k \in K} \delta_d \omega_d \beta_b \alpha_{b,k}; \sum f_i P_{x,y} \right\} \tag{8.6}$$

Subject to:

$$\begin{aligned} \vec{y}_{i,PC} &\geq \vec{x}_{i,PC} \\ \vec{w}_{i,FC} &\geq \vec{u}_{i,FC} \end{aligned} \tag{8.7}$$

The purity constraints for EL and water were defined as 99.5 mol% and 99.5 mol%, whereas the molar flow rate was set at least 99.5 kmol/hour for both EL and water in their respective streams.

The decision variables used for the optimization of the RD processes are shown in Table 8.2. These variables consist of a combination of discrete and continuous variables.

The results obtained by Vázquez-Castillo et al. (2019) are shown in the Pareto front charts of Fig. 8.5. The utopian point methodology was used to select the solution with the best tradeoff between the objectives. Their results indicate that the TCRD process has the lowest energy use (1.667 MJ/kg EL) with major energy savings (9.6%–54.3% lower than other RD processes), additionally, it reduced environmental impact around 5.7%–51% lower ECO 99 index value and it shows similar process safety (less than 2% difference as compared to other RD processes considered). Thus, the TCRD process is suggested as the best process alternative to produce EL. Their results contrasted with the results obtained by Novita et al. (2017). The contrasts are explained by the fact that the implementation of a multiobjective optimization algorithm needs some adjustments to the rigorous process simulation: for example, for the THRD process, the withdrawal side stage number and the side molar flow rate in the first separation column are both variables subject to

Table 8.2 Discrete and continuous decision variables for the optimized RD processes.

Decision variables	CRDP		TCRD		RDHI		THRD		PDWC	
	Cont.	Disc.	Cont.	Disc.	Cont.	Disc.	Cont.	Disc.	Cont.	Disc.
Number of stages, RDC		X		X	X	X		X		X
Number of reactive stages		X		X	X	X		X		X
Heat duty of RDC, kW	X								X	
Distillate flow, kmol/h	X		X				X		X	
Diameter of RDC, m	X		X				X		X	
Number of stages, C1		X		X	X	X		X		X
Feed stage, C1		X		X	X	X		X		X
Reflux ratio of C1	X		X				X			
Interlinking flow, kmol/h			X				X		X	
Bottom flow of C1, kmol/h	X		X				X			
Diameter of C1, m	X		X				X		X	
Withdrawal side stage					X	X		X		
Side flow, kmol/h							X			
Number of stages, C2		X		X	X	X		X		X
Feed stage, C2		X		X	X	X		X		X
Reflux ratio of C2							X		X	
Bottom flow, kmol/h	X		X						X	
Heat duty of C2, kW	X		X				X		X	
Diameter of C2, m	X		X						X	
Total number of variables	15		15		13		17		16	

CRDP, conventional reactive distillation process; *PDWC*, dividing wall column; *RDHI*, reactive distillation with heat integration; *TCRD*, thermally coupled reactive distillation; *THRD*, hybrid heat integration–thermally coupling arrangement.

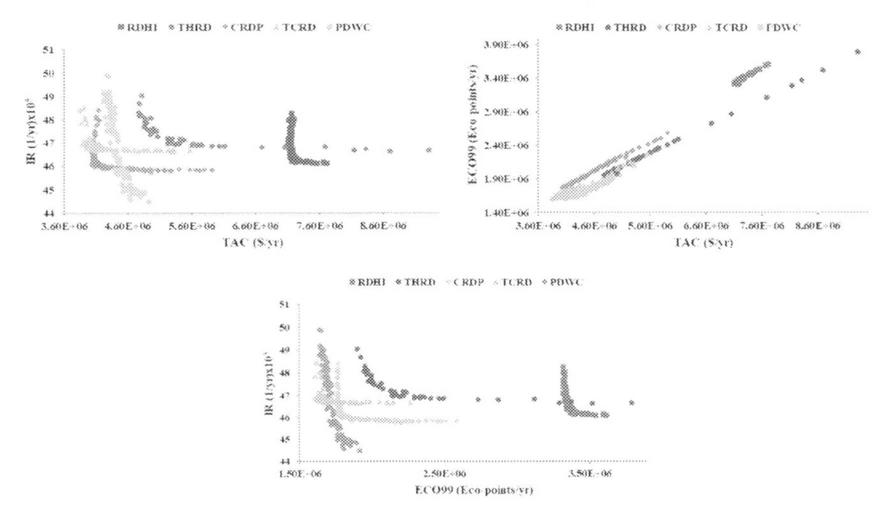

Figure 8.5 Pareto front for the intensified ethyl levulinate reactive distillation.

optimization, whereas an additional constraint was added for the minimum temperature difference (the driving force of 10 K). For this reason, they found that only a part of the condenser energy of C2 was feasible to be reutilized. Based on their results, the TCRD process is the most attractive alternative to be implemented for producing EL. It presented the lowest energy requirements (1.667 MJ/kg EL) and an annual cost of utilities of $30.35 per ton of EL produced, as well as lowest CO_2 emissions (110.4 kg/ton EL) due to thermal coupling.

8.6 Conclusions

This chapter addressed some important features, data, and advances about EL production. As the production of LA has become feasible recently, the production of EL has not been well established. The market data indicates that EL will be an important chemical in the next few years, therefore efforts must be focused on the development of a suitable and profitable process are required. In this sense, the reactive distillation could be a suitable alternative for EL production the first works proposed by Novita et al. (2017) and Vázquez-Castillo et al. (2019) are a good starting point; however, more developments and studies or other alternatives is required.

References

Ahmad, E., Alam, M.I., Pant, K.K., Haider, M.A., 2016. Catalytic and mechanistic insights into the production of ethyl levulinate from biorenewable feedstocks. Green. Chem. 18 (18), 4804–4823.

Corma, A., Iborra, S., Velty, A., 2007. Chemical routes for the transformation of biomass into chemicals. Chem. Rev. 107 (6), 2411–2502.

Expert Market Research (EMR) (2020) https://www.expertmarketresearch.com/reports/ethyl-levulinate-market.

Fang, Z., Qi, X. (Eds.), 2017. Production of Platform Chemicals From Sustainable Resources. Springer, Singapore.

Global Market Insights (GMI) (2020) https://www.gminsights.com/industry-analysis/ethyl-levulinate-market.

Grand View Research (GVR) (2016) https://www.grandviewresearch.com/industry-analysis/ethyl-levulinate-market.

Leal Silva, J.F., Grekin, R., Mariano, A.P., Maciel Filho, R., 2018. Making levulinic acid and ethyl levulinate economically viable: A worldwide technoeconomic and environmental assessment of possible routes. Energy Technol. 6 (4), 613–639.

Mordor Intelligence (2019) https://www.mordorintelligence.com/industry-reports/ethyl-levulinate-market.

Mukherjee, A., Dumont, M.J., Raghavan, V., 2015. Sustainable production of hydroxymethylfurfural and levulinic acid: Challenges and opportunities. Biomass Bioenergy. 72, 143–183.

National Renewable Energy Laboratory (NREL), (2004). https://www.nrel.gov/docs/fy04osti/35523.pdf.

Novita, F.J., Lee, H.Y., Lee, M., 2017. Energy-efficient design of an ethyl levulinate reactive distillation process via a thermally coupled distillation with external heat integration arrangement. Ind. Eng. Chem. Res. 56 (24), 7037–7048.

Ramli, N.A.S., Amin, N.A.S., 2017. Optimization of biomass conversion to levulinic acid in acidic ionic liquid and upgrading of levulinic acid to ethyl levulinate. BioEnergy Res. 10 (1), 50–63.

Tsai C.Y. 2014. Kinetic behavior study on the synthesis of ethyl levulinate over heterogeneous catalyst, Master Thesis, NTUST, Taiwan.

Vázquez-Castillo, J.A., Contreras-Zarazúa, G., Segovia-Hernández, J.G., Kiss, A.A., 2019. Optimally designed reactive distillation processes for eco-efficient production of ethyl levulinate. J. Chem. Technol. & Biotechnol. 94 (7), 2131–2140.

Yadav, G.D., Yadav, A.R., 2014. Synthesis of ethyl levulinate as fuel additives using heterogeneous solid superacidic catalysts: Efficacy and kinetic modeling. Chem. Eng. J. 243, 556–563.

CHAPTER 9
2,3-Butanediol

Contents

9.1 Introduction

Today, because crude oil supplies are becoming scarce, if they are made from bio-based materials, some chemicals commonly produced by synthetic routes are gaining more interest. If its production is based on biomass, 2,3-butanediol (2,3-BD) is a renewable chemical. Since its application covers many industrial sectors, several studies have described 2,3-BD as a very interesting bio-based compound. In order to commercialize biochemicals, high price strategies are necessary. However, since conventionally manufactured biochemicals compete with petrochemicals for similar industrial purposes, market penetration has been difficult. Table 9.1 shows the main physicochemical characteristics of 2,3-BD (Ji et al., 2011).

It is very important to recognize the market for high-value-added areas that are distinct from petrochemical products in this respect. In this respect, 2,3-BD has many possible applications and is a promising industrial chemical (Fig. 9.1). In the chemical, cosmetic, agricultural, pharmaceutical, and food industries, it could be widely used. Conventionally, possible applications of 2,3-BD as a bulk production chemical have centered largely on the chemical and fuel industries. For economic reasons, this scenario has delayed its commercialization using biological production

Improvements in Bio-Based Building Blocks Production Through Process Intensification and Sustainability Concepts
DOI: https://doi.org/10.1016/B978-0-323-89870-6.00011-1

Table 9.1 2,3-BD properties.

Properties	
Chemical formula	$C_4H_{10}O_2$
Molar mass (g/mol)	90.122
Appearance	Colorless liquid
Odor	Odorless
Density (g/mL)	0.987
Melting point (°C)	19
Boiling point (°C)	177
Solubility in water	Miscible
Solubility in other solvents	Soluble in alcohol, ketones, ether
Vapor pressure (hPa at 20°C)	0.23
Acidity (pK_a)	14.9
Heat capacity (kJ/mol)	213.0
Std enthalpy of formation (kJ/mol)	−544.8

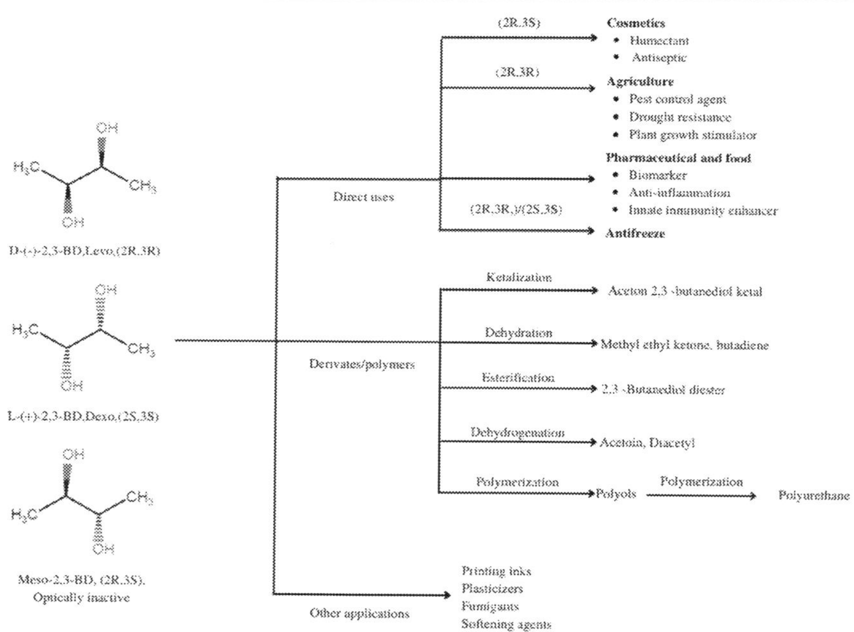

Figure 9.1 Potential 2,3-BD fields of application, including chemical, cosmetics, agricultural, medicinal, and other industries.

processes. However, new areas of application for the use of this compound have recently included the cosmetics, agricultural, pharmaceutical and food industries. Thus, 2,3-BD is again attracting interest. The value increase and market expansion of 2,3-BD could lead to these opportunities.

During World War II, industrial interest in 2,3-BD began because it could be converted through dehydration using chemical catalysis to 1,3-butadiene and then 1,3-butadiene could be used for the production of synthetic rubber (Ji et al., 2011). Although 1,3-butadiene is formed by 1,2-water molecule removal, pinacol rearrangement creates another dehydration product, methyl ethyl ketone (MEK), an important fuel additive as well as an industrial solvent for resins and coatings. High-quality aviation fuel octane can be produced from 2,3-BD (Białkowska, 2016) in combination with MEK and the hydrogenation reaction. It is possible to use (2R,3R)-BD as an antifreeze since its freezing point is less than −30°C. In addition, the polymer synthesizer (Syu, 2001) will act as a monomer. 2,3-BD can also be used in the processing of polyol and polymeric isocyanate, which are polyurethane intermediates, as a novel chain initiator and extender (Paciorek-Sadowska and Czupryński, 2006).

The cosmetics industry now demands affordable, eco-friendly, and natural cosmetic raw materials. In 2013, Loreal, the world's top cosmetics company, announced its latest "Sharing beauty with all" sustainability pledge for 2020 as a representative example (The L'oréal Sustainability Commitment, 2013). A structure with four specific commitments has been created: to invent, produce, live, and build sustainably. As a result, 2,3-BD is one of the raw materials most appropriate for the cosmetics industry's mega-trend shifts. As an ingredient in both cosmetics and personal care products, it has great potential.

It is also expected that 2,3-BD will be used in the agricultural industry for different purposes. The protection of the environment and people is the most important consideration to be considered before using 2,3-BD as an agricultural product. Since it is considered to be a constituent of edible products such as fruits and wines (Jiang and Zhang, 2010), it is recognized as a healthy chemical. We also observed that there was no phytotoxicity impact of 2,3-BD on agricultural crops, such as tomatoes, cucumbers or peppers (Jiang and Zhang, 2010). 2,3-BD can induce major growth promotion and systemic resistance (ISR)-mediated defense effects in Arabidopsis against the bacterial pathogen *Erwinia carotovora* (Ryu et al., 2004). Furthermore, it has been reported that 2,3-BD can induce systemic resistance to *E. carotovora* (Han et al., 2006) at a range of 1 mg to 100 pg/plant in tobacco. It has been documented that 2,3-BD is effective against both fungal and bacterial diseases. Recent studies have shown the effectiveness of 2,3-BD in the treatment of viral diseases. There is no effective antiviral agent currently on the market. Recently, treatment with 2,3-BD has been reported to significantly reduce the incidence of naturally occurring viruses such as Cucumber mosaic virus and Tobacco mosaic virus in comparison with water control, thereby increasing the yield of mature pepper fruits (Kong et al., 2018). As soon as it has been reported that 2,3-BD may cause enhanced innate immunity and result in clearance of damaged liver cells by triggering natural killer cell activity, the pharmaceutical industry might be more interested in this compound (Lai et al., 2012). Based on an earlier study, 2,3-BD has been shown to have anti-inflammatory effects. Endotoxin-induced acute lung damage in rats can be eased (Hsieh et al., 2007). These results show that for human therapeutics, 2,3-BD could theoretically be used. In addition, since 2,3-BD is a natural substance found in naturally prepared or occurring products such as wine, beer, fermented foods, soils, and plants, 2,3-BD can be used as a food additive or health aid on the basis of its effectiveness as an immunity enhancer (Ryu et al., 2004). Furthermore, it is an eco-friendly substance that can be produced by microbial fermentation.

9.2 Production of 2,3-BD from fossil and renewable sources

There are three stereoisomeric forms of 2,3-BD [513—85—9], 2,3-butylene glycol: butanediols, butenediol, and butynediol.

Quite pure butadiene, which has been used in the United States for the manufacture of synthetic rubber, can be obtained by pyrolysis of the diacetate. Butenes from crack gases are the raw material for today.

After removal of butadiene and isobutene from crack gases, a C4 hydrocarbon fraction, called C4 raffinate II, is obtained, which contains approximately 77% butenes and 23% of a mixture of butane and isobutane. By chlorohydrination of this fraction with a solution of chlorine in water and subsequent cyclization of the chlorohydrins with sodium hydroxide, a butene oxide mixture of the following composition is obtained: 55% *trans*-2,3-butene oxide, 30% *cis*-2,3-butene oxide, 15% 1,2-butene oxide.

This mixture hydrolysis (50 bar, 160°C—220°C, reaction enthalpy $H = -42$ kJ/mol) produces a mixture of butanediols separated by vacuum fractionation. A surplus of water must be used to prevent the formation of polyethers during hydrolysis. Butanediol fractionation is simpler than butene oxide fractionation. Meso-2,3-BD is obtained by this reaction sequence from *trans*-2-butene via *trans*-2,3-butene oxide; an analogous racemic mixture of R,R- and S,S-2,3-BD is formed from *cis*-2-butene via *cis*-2,3-butene oxide. See also (Wilson and Lucas, 1936) for a discussion of the stereochemistry of this reaction.

In 1906, Harden and Walpole, who used *Klebsiella pneumoniae*, first documented the efficient microbiological development of that compound about biological pathways. At the time, the most cost-effective and much cheaper than chemical synthesis was their methodology. It was produced on a commercial scale during World War II, when intensive efforts were prompted by a shortage of the strategic compound 1,3-butadiene (an organic intermediate in synthetic rubber production) in the field of diol production (Ji et al., 2011). Over the years, however, interest in microbiological development of 2,3-BD has largely decreased due to the greater availability of cheaper petroleum-based routes. Nowadays, the compound has become costly (USD 1600/ton) due to the volatility of fossil fuel rates, political considerations, and costly chemical catalysts that promote the synthesis of the specific diol structure, as well as the high energy intensity of the process, which has severely detracted from its application potential (Shrivastav et al., 2013).

Concerns regarding the depletion of fossil fuel supplies, CO_2 emissions associated with petroleum products and the accumulation of nondegradable synthetic polymers have stimulated the development of environmentally friendly chemical processes. A solution proposed by the emerging field of white biotechnology is to produce, by microbial fermentation, chemical building blocks. Based on biological catalysts, this constitutes a much safer and more environmentally sustainable technology. Under mild process conditions, their high specificity and catalytic activity will increase the cost-effectiveness of the process, including reducing the number of steps, reducing energy consumption, and reducing both the quantity and the harmfulness of waste products.

Mixed-acid fermentation accompanies the biosynthesis of 2,3-BD in bacterial cells (Fig. 9.2), which is typical of most members of the

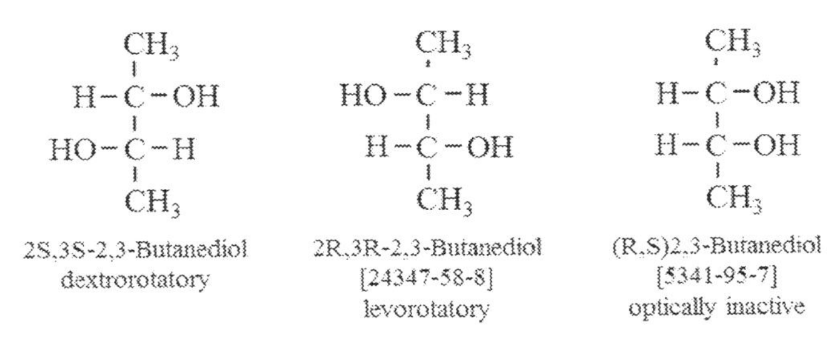

Figure 9.2 2,3-BD stereoisomers.

enterobacteriaceae family, as well as some other aerobic and anaerobic bacteria. This process contributes to the release of acidic compounds, reducing the environmental pH of the microorganisms that perform this type of fermentation gradually.

So far, the 2.3-BD metabolic role has not been determined. The development of 2,3-BD stimulated by reduced pH, however, is believed to prevent intracellular acidification as it decreases the pool of acidic compounds in favor of neutral ones (Ji et al., 2011). The marked activation of butanediol pathway enzymes observed in the presence of acetic acid is also confirmed by this assumption (Zeng et al., 1990).

Does 2,3-BD production also lead to the maintenance of a suitable Nicotinamide adenine dinucleotide (NAD)/Nicotinamide Adenine Dinucleotide (NAD) + Hydrogen (NADH) proportion in the cells, as it is involved in NAD regeneration? (Celińska and Grajek, 2009) developed during glycolysis. 2,3-BD is often used in some microorganisms to store carbon and energy in the stationary process, following the depletion of resources accessible.

9.2.1 Microorganisms useful in the production of 2,3-BD

There are bacteria in the largest and best-studied community of 2,3-BD producers and some other yeasts and microalgae producers are presented in Table 9.2. The application of alcohol synthesis of natural microbiological producers does however not make this method in itself competitive with chemical synthesis from fossil materials. Therefore, as mentioned below, most of the above strains are subject to genetic modification by metabolic engineering.

Currently, the most efficient 2,3-BD producers are the strains *K. pneumoniae*, *Klebsiella oxytoca*, and *Serratia marcescens* (Ji et al., 2011). In glucose-based fed-batch cultures, the first one can synthesize 2,3-BD upto approx. 150 g/L. Genetically modified (GM) versions of the other two strains generate similar quantities of the compound, that is, 142.5 and 152 g/L, utilizing glucose and molasses as a source of sugar in fed-batch cultures, respectively (Ji et al., 2010; Zhang et al., 2010). All these findings show that *Klebsiella* spp. is still considered to be the most promising 2,3-BD manufacturers, with a relatively fast-growing culture and poor nutrients. In addition, not only glucose but other sugars can be converted into the target diol in the cereal hydrolysates and various lignocellulose substances as well as glycerol (Garg and Jain, 1995; Petrov and Petrova, 2009). The synthesis of sugar beets by certain bacteria is suitable, including sugarcane beets

Table 9.2 Microorganisms that produce 2,3-BD

Group of microorganisms	Scientific nomenclature
Bacteria, cyanobacteria	*K. pneumoniae, K. oxytoca, K. terrigena, B. licheniformis, B. amyloliquefaciens, S. marcescens* *B. subtilis, B. stearothermophilus, B. cereus, E. aerogenes, P. polymyxa, P. putida* *A. hydrophilia, A. aerogenes, B. brevis, C. glutamicum, L. brevis* *L. casei, L. helveticus, L. platarum, L. lactis* *L. mesenteroides, P. pentosauces, R. platicola, M. morganii, Pantoea* sp., *Se. Plymuthica* *C. autoethanogenum, C. ljungdahlii, C. ragsdalei, P. chlororaphis 06, E. coli* *Synechocystis* sp., *S. elongates*
Yeast	*S. cerevisiae, K. apiculata, Z. bailii, S. ludwigii, H. uvarum*
Marine microalgae	*C. perigranulata*

K. pneumoniae, K. oxytoca, K. terrigena.- Klebsiella for all K's; B. subtilis, B. stearothermophilus, B. cereus.- Bacillus for all B's; E. aerogenes.- Escherichia aerogenes; P. polymyxa.- Peanibacillus polymyza; P. putida.-Pseudomonas putida; C. glutamicum.- Corynebacterium glutamicum; A. hydrophilia.- Aeromonas hydrophilia; A. aerogenes.- Aerobacter aerogenes; C. glutamicum.- Corynebacterium glutamicum; L. brevis.- Lactobacillus brevis and Lactobacillus for all L's; P. pentosauces.- Pediococcus pentosaceus; R. platicola.-Raoultella planticola; M. morganii.-Morganella morganii; Se. Plymuthica.-Serratia plymuthica; C. autoethanogenum, C. ljungdahlii, C. ragsdalei.- Clostridium for all C's.; P. chlororaphis 06.- Pseudomonas chlororaphis 06; E. coli.- Escherichia coli; S. elongates.- Saccharomyces elongates; S. cerevisiae.- Saccharomyces cerevisiae; K. apiculata.-Klebsiella apiculate; Z. bailii.- Zygosaccharomyces bailii; S. ludwigii.- Saccharomycodes ludwigii; H. uvarum.- Hanseniaspora uvarum; C. perigranulata.-Chlamydomonas perigranulata.

(Afschar et al., 1991) and sugarcane molasses (Dai et al., 2015). Fig. 9.3 shows the metabolic pathways starting from glucose.

However, despite these benefits wild *Klebsiella* strains, due to their pathogenic properties, cannot be implemented on an industrial scale. There are attempts to genetically modify *Klebsiella* mutants that play a key role, in both survival, and pathology of bacteria without any virulence factors like lipopolysaccharides (LPS), polysacchoid capsules, and fimbrial adhesins (Shrivastav et al., 2013). Jung et al. (2013b) use site-specific techniques of recombination to eliminate some other producers of 2,3-BD virulence factors, namely, *Klebsiella* and *Klebsiella Pneumoniae*. They were able to interrupt glucosyltransferase by deletion mutation, which is necessary for synthesis of outer core LPS, in the wabG gene encoding. There were no external core LPS for the wabG mutant strains, which lost their

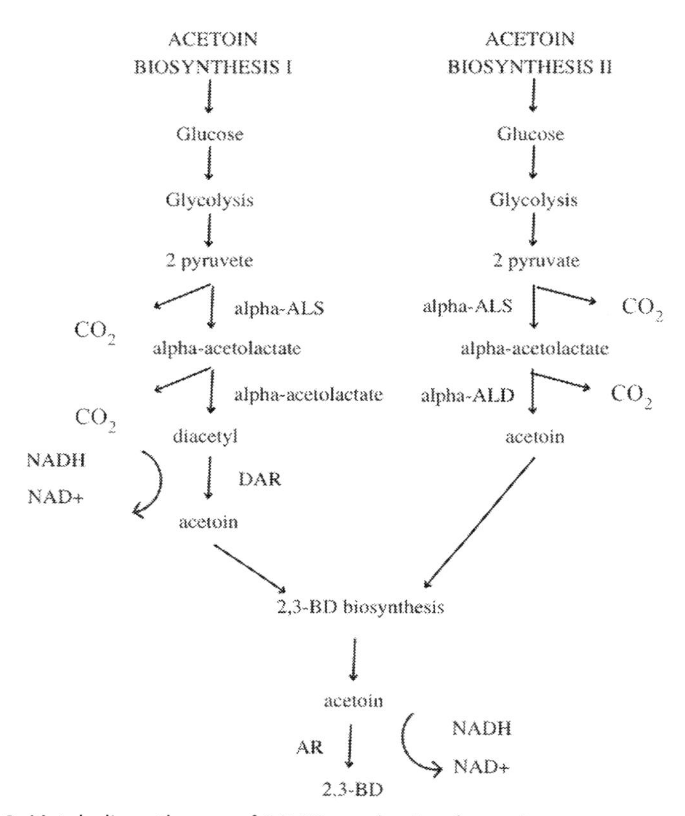

Figure 9.3 Metabolic pathways of 2,3-BD production from glucose.

capsular structure, but their capacity for producing 2.3-BD was roughly reduced. 30% of the wild strains compared (31.27 g/L). Importantly, the bacterial growth rate was not affected by the mutation. Their findings were, according to Jung et al. (2013b), promising as a step towards the development, for commercial uses, of nonpathogenic *Klebsiella* mutants.

Attractive alternative producers of 2,3-BD are also some bacillus strains, which are nonpathogenic and exhibit satisfactory fermentative capability. The most interesting 2,3-BD producer of that genus is *Bacillus licheniformis* DSM 8785, which generated approx. 145 g/L of 2,3-BD in fed-batch culture on a glucose-containing medium (Jurchescu et al., 2013). Similar production levels (133 g/L) were obtained for the mutant *Bacillus amyloliquefaciens* B10−127 (Yang et al., 2013).

Several microorganisms can synthesize stereoisomers of 2,3-BD, but a mixture of two isomers is known in most strains. *Klebsiella* strains develop

L-2,3-BD and meso-2,3-BD and D-(−)-2,3-BD and the meso-2,3-BD are generated by members of the genre *Bacillus*. The choice of the manufacturer, therefore, depends to a large extent on the intended use of alcohol. *Klebsiella* spp. When 2,3-BD is to be used as a basic industrial chemical without regard to its isomeric shape. The commercial development will be the microorganisms of choice. However, if the desired product is a 2,3-BD chiral, the desired product is *Paenibacillus polymyxa*, which generates pure isomer (2R,3R)-2,3-BD (Ji et al., 2011). Thanks to expanded knowledge about molecular mechanisms which support 2,3-BD biosynthesis by different microorganisms and advanced engineering, mutants that generate a certain optically pure type of alcoholic beverage can be created (Fu et al., 2016; Qi et al., 2014; Ui et al., 2004).

Recently, the achievements of synthetic biology have given much hope for engineering photosynthetic bacteria, which would be capable of converting assimilated CO_2—into desirable alcohols, including 2,3-BD. While cyanobacteria have the biochemical machinery required to fix CO_2, they lack the critical components to generate fuels and chemicals efficiently. Thus, it would seem a good idea to construct novel systems by assembling the components and control systems into new combinations. Savakis et al. (2013), using *Synechocystis* spp., defined such an approach. The heterologous catabelic pathway of 2,3-BD is expressed in PCC6803. Enteric and lactic acid bacteria. On the 12th day of culture, the highest diol obtained for GM synechocystis was 0.43 g/L and around 70 times the starting value. Oliver et al. (2013), constructing a 2.3-BD course in other cyanobacterium cells, *Synechococcus elongatus* PCC7942, showed also encouraging results. The engineered line reached a production rate of 9.847 μg/L hour and a final titer of 2.38 g, with continuous production lasting 21 days during continuous CO_2 and light production. But, as opposed to engineered strains of heterotrophic bacteria, however, the amounts of diol synthesized by GM cyano-bacteria remain very tiny. Further work is nevertheless hoped to change the photosynthesizing bacteria to allow new CO_2-reduction manufacturing technologies. This is important because atmospheric CO_2 levels in the last 150 years have risen approximately 25% (Oliver et al., 2013).

9.3 Raw material for 2,3-BD production

Various raw materials were measured for the industrial production of 2,3-BD based upon costs and sustainability (Fig. 9.3). As the costs of raw material represent a significant proportion of the overall cost of production, cheap and plentiful biomass was investigated. The use of waste products and surplus biomass is an optimal option, because of the resistance associated with food supplies in the industrial production of biochemicals. Among the biomasses used, whey, sugarcane, manga, Jerusalem artichoke and crude glycerol (Białkowska, 2016; Gao et al., 2010; Petrov and Petrova, 2009; Wang et al., 2012) are relatively inexpensive and plentiful substrates. A biorefinery is also an attractive alternative to lignocellulosic biomass, as it is an abundant source worldwide. It is also available at a low cost on a sustainable basis.

9.3.1 Nonrenewable raw materials

Sugarcane industry by-products, for example molasses, contains various nutrients, including sucrose, minerals, organic compounds, and vitamins. Due to its low price and rich sugar composition, sugarcane molasses have been acknowledged as an appropriate carbon source for fermentation (Akaraonye et al., 2012). Mixed sugars, however, are used in molasses less effectively than pure sucrose because the pathways to use substrate are subject to carbon catabolite repression (CCR) (Yang et al., 2017). To resolve this obstacle, it is therefore important to engineer the regulatory framework in fed-batch fermentation with sugarcane molasses as the key carbon source for sucrose regulators (ScrR) encoding gene-deleted E, for instance, the high concentration of 2,3-BD exceeding 96 g/L was achieved (M.Y. Jung et al., 2013a) strain. The raw sugar (sucrose) derived from sugarcane containing a limited impurity level was also validated as a promising source of CO_2 in various microorganisms for efficient 2,3-BD processing, but more costly than molasses (Song et al., 2018; Xin et al., 2016).

Cassava is a relatively cheap and plentiful starch crop that is widely grown in the tropics and subtropics. Due to clear seeding and harvest methods, this is available throughout the year. Furthermore it lacks competition as regards the price of cassava starch in the food industry compared to maize starch (Pervez et al., 2014). It does not have significant applications. It is considered one of the most productive carbohydrate crops because cassava roots have a high content of starch. It can be used in a variety of ways such as mango starch, manioc powder, cassava baggasse, cassava chips, and fresh manioc root. A newly isolated E in one study. For 2,3-BD output, the *cloace* strain was used using carbon powder cassava (Wang et al., 2012).

Under optimum conditions, after 47 h of simultaneous saccharification and fermentation (SSF) (Wang et al., 2012), 93.9 g/L of 2,3-BD was generated. The Jerusalem artichoke is a cheap and plentiful raw material. Fresh Jerusalem artichoke tubers contain about 20% carbohydrates, more than 70% of which are inulin (Szambelan et al., 2004).

Inulin consists of several units of fructose with glucose termination. Fructose and glucose can be hydrolyzed by inulinase. 2,3-BD by K output. Jerusalema Artichoke Tuber *Pneumoniae* has succeeded in generating the fed-batch SSF process with a capacity of 91.6 g/L (Sun et al., 2009) after 40 h. Inulin hydrolysate has also been used for 2,3-BD production by thermophilic *B. licheniformis* strain, with a titer of 103 g/L and a high productivity of 3.4 g/L/h through a fed-batch SSF process.

A strain named *P. polymyxa* was utilized for the direct use of the articulated tuber of Jerusalem for 2,3-BD development without inulinase supplement. This strain secrets inulinase naturally. Then 36.92 g/L (2R,3R)-BD with 98% purity was manufactured through one-stage batch fermentation using raw inulin as a substratum (Gao et al., 2010). Whey is a milk industry by-product. It is made up of around 5% lactose and 1% protein. The use of whey for 2,3-BD fermentation was investigated by several groups in view of its readily accessibility in many countries. However, whey sugars are of low concentrations. In effect, fermentation efficiency using whey as a substratum is comparatively poor compared to traditional substrates such as starch and sugar. In order to address this issue of low productivity, cell immobilization technologies have been developed (Champluvier et al., 1989). Nevertheless, process efficiencies for industrial applications do need to be improved. Glycerol is a by-product of the fermentation of ethanol, fat saponification and processing of biodiesel. It is also a promising medium for the manufacturing of 2,3-BD (Ahn et al., 2011) in industrial production. Notice that the world market is projected to generate an increasing surplus of glycerol in the short term because there is a growing demand for biodiesel and a total weight of almost 10% of biodiesel-derived cruel glycerol (da Silva et al., 2009). The only source of carbon of the metabolically engineered K is biodiesel-derived crude glycerol. The delete of the pduC (the encoding subunit of glycerol dehydratease) and the ldhA genes for by-products including 1,3-propanediol and lactate (Cho et al., 2015) is the *oxytoca* M1. Double mutant strain could produce 131.5 g/L of 2,3-BD of crude glycerol in a fed-batch fermentation, respectively with yields and productivity of 0.44 g/g · crude glycerol and 0.84 g/L/hour.

9.3.2 Renewable raw materials

Lignocellulose comprises various monomers of sugar closely connected to the aromatic polymer (cellulose and hemicellulose) (lignin). This compact structure makes lignocellulosic biomass difficult to apply directly. Preprocessing of biomass hydrolysis to free sugar is typically needed before lignocellulosic biomass is applied in high-temperature conditions with acid or ammonia solutions. Acid hydrolysis normally produces various toxic derivatives, including furfural and 5-hydroxymethylfurfural, which can inhibit microorganism growth and impede the performance of production (Thomsen et al., 2009). Without competing with the food industry, lignocellulosic biomass is cost-effective and plentiful, making it a valuable feedstock in biorefineries. Various lignocellulosic biomasses of agriculture residues and wood such as maize cob, maize stover and apple pomace have been used in the development of 2,3-BD on an industrial scale with lignocellulosic biomasses (Thomsen et al., 2009) and have also been verified. Since lignocellulose hydrolysis generates mixtures of glucose and xylose-containing sugars (Sheehan and Himmel, 1999), strain engineering is generally important for the efficient use and conversion of the sugars into 2.3-BD.

Corn stover hydrolysate was tested as an industry substratum that is extremely promising. There have been some demonstration studies. The simultaneous deletion of glucose transporter-encoding ptsG gene and overexpression of galactose encoding GalP in E remove carbon catabolite substitution in one sample. Simultaneous use of glucose and xylose *cloace* strain SDM. In addition, 119.5 g/L of (2R,3R)-BD were produced using corn stover material as raw material after 52 hours of fermentation with a productivity of 2.5 g/L (Thomsen et al., 2009) by means of additional engineering strategies for minimizing by-product formation and improving the specific output of (2R,3R)-BD. Another example was the fed-batch fermentation of the newly isolated strain *B. licheniformis* X10, with high fermentation inhibitor tolerance, such as furfure, vanillin, formic acid, and acetic acid, and the result was that corn storage hydrolysate with a production capacity of 2,1 g/L/hour was produced at 74 g/l (2,3-BD).

Corn molasses are also one of the candidates for the industrial processing of 2.3-BD as waste product resulting from xylitol production. For instance, K generated 78.9 g/L of 2,3-BD. Power of *pneumoniae* after 61 hours of fermentation fed-batch (Wang et al., 2010). Apple pomace is an abundant waste after the juice is squeezed. It has been used more and more in biotechnology applications. Apple pomace hydrolysate is currently used in the processing of microbial 2,3-BDs. For instance, it was used to produce 113 g/L 2,3-BD. NCIMB 8059 strain *licheniformis* (Białkowska et al., 2015). Also investigated as raw materials for the production of 2,3-BD were other low-cost biomasses such as empty palm fruit bunkers and marine algae biomass (Kang et al., 2015; Mazumdar et al., 2013). Depending on the production host used to test and evaluate different forms of biomass, the usable biomass may be improved and evaluated in the future.

9.4 Process intensification (PI) in 2,3-BD production

PI is the development of innovative devices and technologies that provide significant improvement in chemical manufacturing and processing, a significant reduction in the amount of equipment, energy consumption, or waste production, and eventually lead to more cost-efficient, safer, and sustainable technologies (Stankiewicz and Moulijn, 2000).

In the particular case of 2,3-BD, PI could be applied in generating various technologies that improve performance. This type of technology is relatively common in other biomolecules, for example, ethanol, butanol, and so on.

However, the study focused on PI has been mostly focused on the genetic modification of various species involved in the production of 2,3-BD to increase yields. While research efforts on PI for 2,3-BD production appear to be few, the interest in 2,3-BD is driven by its large application potential, expressed in a 4%–7% annual increase in output (Shrivastav et al., 2013). The figures relate to the chemical processing of 2,3-BD, which is high in energy and demands a costly catalyst, but which is the only process used in the production of this alcohol on an industrial scale to date. While other microbiological techniques are not economical because of their low performance, they can be cheaper and more environmentally friendly. The current task is therefore to improve the effectiveness and economic performance of new solutions. The subject literature shows numerous approaches to solving the problem, focusing mainly on the genetic engineering of strains generating 2,3-BD with a high yield, using low-cost alternative substrates, and optimizing process conditions. The following methods are illustrated.

DNA recombination technology to overexpress the gene(s) involved in diol biosynthesis in host cells and/or the genes responsible for competitive pathways is used to alter genetic modification by 2,3-BD producers in order to avoid biosynthesis of undesirable fermentation products, such as ethanols, acetoin, lactic acid, and acetate. Satisfactory results can generally be obtained via methods of metabolic engineering that accompany the expression of the objective gene(s) with the suppression of the competing gene(s). The majority of reports in this field relate to *Klebsiella bacteria*, whose wild strains produce 2,3-BD (Table 9.2). For example, the aldA gene encoding the dehydrogenase of the ME-UD-3K mutant has been disrupted by Ji et al. (2010). Oxytoca produced 130 g/L 2.3-BD with a productivity of 1.63 g/L 9 hours with a yield of 0.48 g/g glucose by inserting a marker for the tetracycline resistance. The similar development rates of 1,4-BD from crude glycerol using the mutant K have been obtained by Cho et al. (2015). *Oxytoca* M1, 2.3-BD was provided by the adjusted strain in a 131.5 g/L concentration of 0.84 g/L 9 H and 0.44 g/g crude glycerol. *Oxytoca* M1 K. In this case, 2.3-BD is produced at a concentration, rates and overall and basic productivities of 117,4 g/L, 0,49 g/G, 1,2 g/L, 9 hours and 27,2 g/g cell dry weight (CDW) respectively, during the fermentation at fed-batch. *Oxytoca* genome was further updated by Jantama et al. (2015), which constructed a KMS 005−73 T mutant.

Moreover, the *Oxytoca* species Klebsiella is being updated with intense effort. Air conditioning. *Pneumoniae*. Guo et al. (2014) produced KG1 inactivated lactate dehydrogenase (ldhA), acetaldehyde dehydrogenase (adh E) and phosphotransacetylase (pta) genes that were responsible for the maximum 2,3-BD output of glucose in fed batches. 100 g/L 2,3-BD at 0.49 g/g yield.

Economic microbiological development of 2,3-BD is not restricted to *Klebsiella* strains and other bacterial species, especially in the enterobacter, Escherichia, and bacillus genera, are involved. For example, Jung et al. (2012) found a substantial rise in the 2,3-BD biosynthesis levels of the deletion of a gene (lactate dehydrogenase) from *Enterobacter aerogenes*. The fed-batch fermentation of the alcohol titer in 54 hours at 69.5 g/L was 27.4% above the parental strain. The mutant produced 118.05 g/L of 2,3-BD over 54 hours of fed-batch fermentation with further optimization of the medium and aerating conditions. The genetically well-characterized *Escherichia coli* has also gained increased interest, which in simple media develops easily and quickly. *E. coli* showed the best synthesis ability for 2,3-BD incorporated with a gene cluster. *Cloacae* producing a fed-batch culture of 2,3-BD in the sum of 73.8 g/L.

9.5 PI in 2,3-BD recovery

Recovery to make the process economically competitive by microbiologic 2,3-BD should boost its performance, and an efficient and cheap method of separating 2,3-BD from the fermentation broth should also be introduced. It is particularly difficult to recover 2,3-BD from the culture medium due to its high boiling point (180°C−18°C), great water affinity and dissolved and solid components of the fermentation mash, which can thus pose a key barrier to marketing. As stated in Xiu and Zeng reviews by Xiu and Zeng (2008) and Ji et al. (2011), a large number of 2,3-BD separation techniques have been tested to date, and have posed some challenges and limitations on the recovery, yield, and energy consumption of various downstream technologies. Pretreatment and a high energy supply, which raise prices, are the key approaches addressed. Some emerging processes of efficient 2,3-BD recovery, including aqueous, two-phase (Jiang et al., 2009) removal of solvent (Shao and Kumar, 2009), in-situ recovery (Anvari and Khayati, 2009) and reactive extraction, have been developed over the previous decade (Li et al., 2013). These processes increase energy consumption and product recovery efficiency and can be useful for increasing the fermentation rate of 2,3 BD. The techno-economic assessment of the whole bioprocess of 2,3-BD output from renewable resources, including the separation point, was verified by Koutinas et al. (2016). Downstream separation of 2,3-BD was based on reactive 2,3-BD extraction using aldehyde to turn the diol into an immiscible acetal of water for each step of fermentation. It can also be noted that the costs of a commodity restoration facility downstream are considerably lower in every case than the costs of a fermentation facility.

To date, few options are available to purify 2,3-BD, for example, the simulated moving bed (SMB) is already used in the LanzaTech technology (Xian, 2015), however, this technology has many disadvantages, alternating pump loads reduce lifetime during operation and which cause faults during operation, in addition to costly material used in SMB (Harvianto et al., 2018). Some separation options have also been tested in the ongoing research for alternatives; for instance, the pervaporation process using polydimethylsiloxane and ZSM-5 zeolite particles is reasonably effective on an experimental scale. However, some obstacles are facing and the membrane activity decreases over time. The membrane often has significant disturbances to functionality including membrane swelling, due to the relative complexity of the fermentation cooler (Koutinas et al., 2016). Other solutions, such as reactive extraction, have shown promising results, but the chemicals used in the process produce high levels of corrosion and environmental problems; in this way, anticorrosion devices must be used to prevent corrosion. In addition, the method is not very mature to be applied on a wide scale (Koutinas et al., 2016; Li et al., 2012).

Recently, Sánchez-Ramírez et al. (2019) presented the synthesis and design of certain alternatives to purifying 2,3-BD based on distillation, which is synthesized in such a way that results are thermally coupled, thermodynamically equivalent and enhanced sequences. In addition, a dividing wall column (DWC) scheme was also planned and optimized to compare its efficiency with other schemes. Considering the three alternatives, they were designed and optimized.

As objective functions, it was considered the total annual cost (TAC) as an economic index, the eco indicator 99 as an environmental index and the inherent risk (IR) of the process (analyzed as individual risk). In general, the improved alternative offered a 15% reduction in the TAC and a 14% reduction in the environmental effects. In addition, the same alternative posed the lowest IR with a reduction of around 50% relative to the reference alternative.

In general terms, the technique for predicting new alternative schemes is based on the implementation of thermal couplings, transposition and elimination of column section. As a product of fermentation, Diol concentrations are 150 and 10 g/L of 2,3-BD and acetoin, respectively (see Table 9.3). Additionally, Fig. 9.4 shows all the intensified alternatives derived from two conventional schemes.

Table 9.3 Feed characterization.

Mass fraction			Vapor fraction	Flowrate (kg/h)	Temperature (K)	Reference
Water	Acetoin	2,3-BD				
0.841	0.009	0.149	0	73,170	298	Ma et al. (2009)

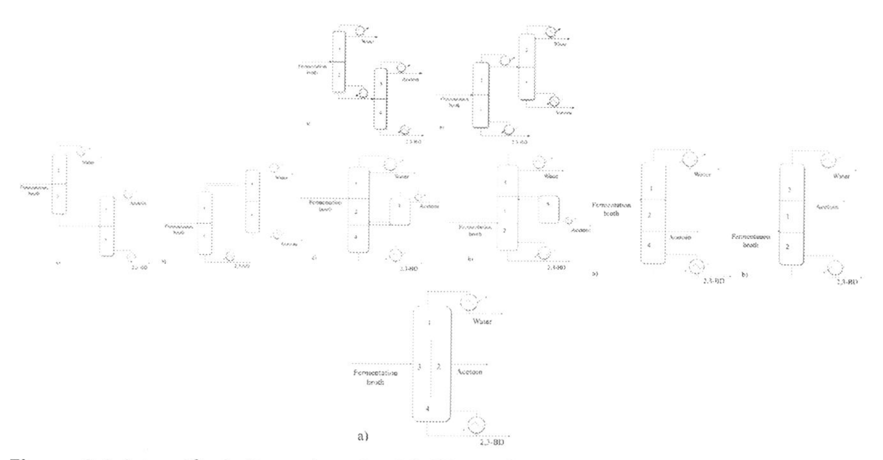

Figure 9.4 Intensified alternatives for 2,3-BD purification.

Table 9.4 Objective function values for all configurations.

Objective function	Direct figure 9.4–1 (a)	Indirect figure 9.4–1 (b)	Thermally coupled figure 9.4–2 (a)	Thermodynamic equivalent figure 9.4–3(a)	Intensified figure 9.4–4 (a)	DWC (DIviding Wall Column) figure 9.4–5
TAC ($/year)	35032419	51155609	31360313	313055124	30536031	231268638
EI99 (points/year)	14328558	22627903	12559191	12557857	12407199	97659879
IR (probability/year)	0.0006686	0.0006663	0.0006795	0.0006684	0.00033411	0.0003661

Once the optimization test was implemented via a synthesis process to the proposed alternatives, the initial results revealed that the direct sequence was better than the indirect scheme between the two traditional alternatives. The direct sequence showed the best performance index (economical, environmental and safety) compared to all alternatives, with intensified alternatives that purify the feed stream in a single column. In contrast with the direct sequences, economic savings could have been achieved by approximately 15%, environmental effect mitigation by 14% and IR savings of approximately 50%. Table 9.4 show the objective functions obtained after the optimization process.

The strong dependency of this index on the size of the equipment, the number of internal flows and the form of chemicals to be segregated was evident about the IR assessment. In other words, because this objective function is directly in competition with the other function, the algorithm pursued reflux ratio parameters, equipment sizing, interconnection flows, etc. through the optimization process to decrease the objective function (Table 9.5).

It is clear that there are few studies for large-scale production of 2,3-BD. In this sense, there is a great niche of opportunity for research and applied engineering. On the other hand, the dynamics of this type of process should also be studied, in order to visualize the real feasibility of implementing any type of new technology.

Table 9.5 Design parameters and performance indexes for the intensified scheme.

	C1
Number of theoretical stages	87
Reflux ratio	0.333
Feed stage	42
Column diameter (m)	0.675
Column height (m)	51.8
Overall efficiency	0.6838
Operative pressure (kPa)	101.353
Distillate flowrate (kmol/h)	3416.07
Side stage	49
Side flow (kmol/h)	8.155
Condenser duty (kW)	51,507
Reboiler duty (kW)	61,457
TAC ($/year)	30,536,031
Eco-Ind (points/year)	12,407,199
IR (points/year)	0.00033411

9.6 Conclusions

As shown, 2,3-BD is a compound that can be used as a platform for various products of domestic and industrial interest. Although the limitations for biological production are clear, recently several studies have been generated to increase productivity and performance.

The sense of the intensification of processes to improve in a global way the production of 2,3-BD, has been very oriented to the understanding of the metabolism of the microorganisms. As a result of the research works, it has been possible to create GM strains that have been able to increase the efficiency in the production of 2,3-BD; additionally, it has been generated a better understanding of the metabolic routes involved in the production of 2,3-BD. Likewise, it has allowed a particular understanding of the enzymatic process in order to obtain the correct relations of stereoisomers.

Recently, the area of bioinformatics, which in silico allows improving the creation of metabolic networks for the production of 2,3-BD based on genomic annotations and databases of metabolic pathways.

On the other hand, it is clear the lack of applied engineering studies, which through physical/mechanical improvements allow improving the productivity of the reactive and separation process. Particularly, the approaches focused on the topological intensification of the process have been few, compared to those observed in other productive biological processes. The use of intensified technologies that try, in an in-situ way, to increase productivity has been relatively few. Similarly, intensified schemes oriented to the downstream process have been relatively few. In that sense, the use of intensified process routes in continuous growth seems to be the key to make the production process to obtain 2,3-BD economically competitive in the short term; for example, the use of hybrid schemes, adsorbents with high separation capacity, membranes with high selectivity, and so on, whose only purpose is the energetic and cost reduction of the process to be economically competitive. Finally, it is also necessary a robust control study to study the real feasibility of any kind of alternative.

References

Afschar, A.S., Bellgardt, K.H., Vaz Rossell, C.E., Czok, A., Schaller, K., 1991. The production of 2,3-butanediol by fermentation of high test molasses. Appl. Microbiol. Biotechnol. 34, 582–585.

Ahn, J.H., Sang, B.I., Um, Y., 2011. Butanol production from thin stillage using *Clostridium pasteurianum*. Bioresour. Technol. 102, 4934–4937.

Akaraonye, E., Moreno, C., Knowles, J.C., Keshavarz, T., Roy, I., 2012. Poly(3-hydroxybutyrate) production by *Bacillus cereus* SPV using sugarcane molasses as the main carbon source. Biotechnol. J. 7, 293–303.

Anvari, M., Khayati, G., 2009. In situ recovery of 2,3-butanediol from fermentation by liquid–liquid extraction. J. Ind. Microbiol. Biotechnol. 36, 313–317.

Białkowska, A.M., 2016. Strategies for efficient and economical 2,3-butanediol production: new trends in this field. World J. Microbiol. Biotechnol. 32, 1–14.

Białkowska, A.M., Gromek, E., Krysiak, J., Sikora, B., Kalinowska, H., Jędrzejczak-Krzepkowska, M., et al., 2015. Application of enzymatic apple pomace hydrolysate to production of 2,3-butanediol by alkaliphilic *Bacillus licheniformis* NCIMB 8059. J. Ind. Microbiol. Biotechnol. 42, 1609–1621.

Celińska, E., Grajek, W., 2009. Biotechnological production of 2,3-butanediol — Current state and prospects. Biotechnol. Adv. 27, 715–725.

Champluvier, B., Francart, B., Rouxhet, P.G., 1989. Co-immobilization by adhesion of β-galactosidase in nonviable cells of *Kluyveromyces lactis* with *Klebsiella oxytoca*: conversion of lactose into 2, 3-butanediol. Biotechnol. Bioeng. 34, 844–853.

Cho, S., Kim, T., Woo, H.M., Kim, Y., Lee, J., Um, Y., 2015. High production of 2,3-butanediol from biodiesel-derived crude glycerol by metabolically engineered *Klebsiella oxytoca* M1. Biotechnol. Biofuels 8, 1–12.

da Silva, G.P., Mack, M., Contiero, J., 2009. Glycerol: a promising and abundant carbon source for industrial microbiology. Biotechnol. Adv. 27, 30–39.

Dai, J.Y., Zhao, P., Cheng, X.L., Xiu, Z.L., 2015. Enhanced production of 2,3-butanediol from sugarcane molasses. Appl. Biochem. Biotechnol. 175, 3014–3024. Available from: https://doi.org/10.1007/s12010-015-1481-x.

Fu, J., Huo, G., Feng, L., Mao, Y., Wang, Z., Ma, H., et al., 2016. Metabolic engineering of *Bacillus subtilis* for chiral pure meso-2,3-butanediol production. Biotechnol. Biofuels 9, 1–14.

Gao, J., Xu, H., Li, Q.J., Feng, X.H., Li, S., 2010. Optimization of medium for one-step fermentation of inulin extract from Jerusalem artichoke tubers using *Paenibacillus polymyxa* ZJ-9 to produce R,R-2,3-butanediol. Bioresour. Technol. 101, 7076–7082.

Garg, S.K.K., Jain, A., 1995. Fermentative production of 2, 3-butanediol: a review. Bioresour. Technol. 51, 103–109.

Guo, X., Cao, C., Wang, Y., Li, C., Wu, M., Chen, Y., et al., 2014. Effect of the inactivation of lactate dehydrogenase, ethanol dehydrogenase, and phosphotransacetylase on 2,3-butanediol production in *Klebsiella pneumoniae* strain. Biotechnol. Biofuels 7, 1–11.

Han, S.H., Lee, S.J., Moon, J.H., Park, K.H., Yang, K.Y., Cho, B.H., et al., 2006. GacS-dependent production of 2R,3R-butanediol by *Pseudomonas chlororaphis* O6 is a major determinant for eliciting systemic resistance against *Erwinia carotovora* but not against *Pseudomonas syringae* pv. tabaci in tobacco. Mol. Plant-Microbe Interact 19, 924–930.

Harvianto, G.R., Haider, J., Hong, J., Van Duc Long, N., Shim, J.J., Cho, M.H., et al., 2018. Purification of 2,3-butanediol from fermentation broth: process development and techno-economic analysis. Biotechnol. Biofuels 11, 1–16.

Hsieh, S.C., Lu, C.C., Horng, Y.T., Soo, P.C., Chang, Y.L., Tsai, Y.H., et al., 2007. The bacterial metabolite 2,3-butanediol ameliorates endotoxin-induced acute lung injury in rats. Microbes Infect. 9, 1402–1409.

Jantama, K., Polyiam, P., Khunnonkwao, P., Chan, S., Sangproo, M., Khor, K., et al., 2015. Efficient reduction of the formation of by-products and improvement of production yield of 2,3-butanediol by a combined deletion of alcohol dehydrogenase, acetate kinase-phosphotransacetylase, and lactate dehydrogenase genes in metabolically engineered *Klebsiella oxytoca* in mineral salts medium. Metabolic Engineering. Elsevier.

Ji, X.J., Huang, H., Zhu, J.G., Ren, L.J., Nie, Z.K., Du, J., et al., 2010. Engineering *Klebsiella oxytoca* for efficient 2,3-butanediol production through insertional inactivation of acetaldehyde dehydrogenase gene. Appl. Microbiol. Biotechnol. 85, 1751−1758.

Ji, X.J., Huang, H., Ouyang, P.K., 2011. Microbial 2,3-butanediol production: a state-of-the-art review. Biotechnol. Adv. 29, 351−364.

Jiang, B., Zhang, Z., 2010. Volatile compounds of young wines from cabernet sauvignon, cabernet gernischet and chardonnay varieties grown in the loess plateau region of China. Molecules 15, 9184−9196.

Jiang, B., Li, Z.G., Dai, J.Y., Zhang, D.J., Xiu, Z.L., 2009. Aqueous two-phase extraction of 2,3-butanediol from fermentation broths using an ethanol/phosphate system. Process Biochem. 44, 112−117.

Jung, M.Y., Ng, C.Y., Song, H., Lee, J., Oh, M.K., 2012. Deletion of lactate dehydrogenase in *Enterobacter aerogenes* to enhance 2,3-butanediol production. Appl. Microbiol. Biotechnol. 95, 461−469.

Jung, M.Y., Park, B.S., Lee, J., Oh, M.K., 2013a. Engineered *Enterobacter aerogenes* for efficient utilization of sugarcane molasses in 2,3-butanediol production. Bioresour. Technol. 139, 21−27.

Jung, S.G., Jang, J.H., Kim, A.Y., Lim, M.C., Kim, B., Lee, J., et al., 2013b. Removal of pathogenic factors from 2,3-butanediol-producing *Klebsiella* species by inactivating virulence-related wabG gene. Appl. Microbiol. Biotechnol. 97, 1997−2007.

Jurchescu, I.M., Hamann, J., Zhou, X., Ortmann, T., Kuenz, A., Prüße, U., et al., 2013. Enhanced 2,3-butanediol production in fed-batch cultures of free and immobilized *Bacillus licheniformis* DSM 8785. Appl. Microbiol. Biotechnol. 97, 6715−6723.

Kang, I.Y., Park, J.M., Hong, W.K., Kim, Y.S., Jung, Y.R., Kim, S.B., et al., 2015. Enhanced production of 2,3-butanediol by a genetically engineered *Bacillus* sp. BRC1 using a hydrolysate of empty palm fruit bunches. Bioprocess Biosyst. Eng. 38, 299−305.

Kong, H.G., Shin, T.S., Kim, T.H., Ryu, C.M., 2018. Stereoisomers of the bacterial volatile compound 2,3-butanediol differently elicit systemic defense responses of pepper against multiple viruses in the field. Front. Plant Sci. 9, 1−13.

Koutinas, A.A., Yepez, B., Kopsahelis, N., Freire, D.M.G., de Castro, A.M., Papanikolaou, S., et al., 2016. Techno-economic evaluation of a complete bioprocess for 2,3-butanediol production from renewable resources. Bioresour. Technol. 204, 55−64.

Lai, H. C., Chang, C. J., Yang, C. H., Hsu, Y. J., Chen, C. C., Lin, C. S., ... Lu, C. C. 2012. Activation of NK cell cytotoxicity by the natural compound 2, 3-butanediol. J. Leukoc. Biol. 92(4), 807−814.

Li, L., Chen, C., Li, K., Wang, Y., Gao, C., Ma, C., et al., 2014. Efficient simultaneous saccharification and fermentation of inulin to 2,3-butanediol by thermophilic *Bacillus licheniformis* ATCC 14580. Appl. Environ. Microbiol. 80, 6458−6464. Available from: https://doi.org/10.1128/AEM.01802-14.

Li, Y., Wu, Y., Zhu, J., Liu, J., 2012. Separation of 2,3-butanediol from fermentation broth by reactive extraction using acetaldehyde-cyclohexane system. Biotechnol. Bioprocess Eng. 17, 337−345.

Li, Y., Zhu, J., Wu, Y., Liu, J., 2013. Reactive extraction of 2,3-butanediol from fermentation broth. Korean J. Chem. Eng. 30, 154−159.

Ma, C., Wang, A., Qin, J., Li, L., Ai, X., Jiang, T., ... Xu, P. 2009. Enhanced 2, 3-butanediol production by Klebsiella pneumoniae SDM. Appl. Microbiol. Biotechnol. 82(1), 49−57.

Mazumdar, S., Lee, J., Oh, M.K., 2013. Microbial production of 2,3 butanediol from seaweed hydrolysate using metabolically engineered *Escherichia coli*. Bioresour. Technol. 136, 329−336.

Oliver, J.W.K., Machado, I.M.P., Yoneda, H., Atsumi, S., 2013. Cyanobacterial conversion of carbon dioxide to 2,3-butanediol. Proc. Natl. Acad. Sci. USA. 110, 1249−1254.

Paciorek-Sadowska, J., Czupryński, B., 2006. New compounds for production of polyurethane foams. J. Appl. Polym. Sci. 102, 5918−5926.

Pervez, S., Aman, A., Iqbal, S., Siddiqui, N.N., Ul Qader, S.A., 2014. Saccharification and liquefaction of cassava starch: an alternative source for the production of bioethanol using amylolytic enzymes by double fermentation process. BMC Biotechnol 14, 1−10. Available from: https://doi.org/10.1186/1472-6750-14-49.

Petrov, K., Petrova, P., 2009. High production of 2,3-butanediol from glycerol by *Klebsiella pneumoniae* G31. Appl. Microbiol. Biotechnol. 84, 659−665.

Qi, G., Kang, Y., Li, L., Xiao, A., Zhang, S., Wen, Z., et al., 2014. Deletion of meso-2,3-butanediol dehydrogenase gene budC for enhanced D-2,3-butanediol production in *Bacillus licheniformis*. Biotechnol. Biofuels 7, 1−12.

Ryu, C.M., Farag, M.A., Hu, C.H., Reddy, M.S., Kloepper, J.W., Paré, P.W., 2004. Bacterial volatiles induce systemic resistance in Arabidopsis. Plant Physiol. 134, 1017−1026.

Sánchez-Ramírez, E., Quiroz-Ramírez, J.J., Hernández, S., Segovia Hernández, J.G., Contreras-Zarazúa, G., Ramírez-Márquez, C., 2019. Synthesis, design and optimization of alternatives to purify 2, 3-butanediol considering economic, environmental and safety issues. Sustain. Prod. Consum. 17, 282−295.

Savakis, P.E., Angermayr, S.A., Hellingwerf, K.J., 2013. Synthesis of 2,3-butanediol by *Synechocystis* sp. PCC6803 via heterologous expression of a catabolic pathway from lactic acid- and enterobacteria. Metab. Eng. 20, 121−130.

Shao, P., Kumar, A., 2009. Recovery of 2,3-butanediol from water by a solvent extraction and pervaporation separation scheme. J. Memb. Sci. 329, 160−168.

Sharing Beauty With All The L'oréal Sustainability Commitment 2013-2020. Closing Report Editorial by Jean-Paul Agon and Alexandra Palt https://www.loreal-finance.com/system/files/2021-03/SBWA_PR_GROUPE_2020_ENG.pdf (visited May 30th, 2021).

Sheehan, J., Himmel, M., 1999. Enzymes, energy, and the environment: a strategic perspective on the United States Department of Energy's Research and Development Activities for Bioethanol. Biotechnol. Prog. 15, 817−827.

Shrivastav, A., Lee, J., Kim, H.Y., Kim, Y.R., 2013. Recent insights in the removal of klebseilla pathogenicity factors for the industrial production of 2,3-butanediol. J. Microbiol. Biotechnol. 23, 885−896.

Song, C.W., Rathnasingh, C., Park, J.M., Lee, J., Song, H., 2018. Isolation and evaluation of *Bacillus* strains for industrial production of 2,3-butanediol. J. Microbiol. Biotechnol. 28, 409−417.

Stankiewicz, A., Moulijn, J.A., 2000. Process intensification: transforming chemical engineering. Chem. Eng. Prog. 96, 22−33.

Sun, L.H., Wang, X.D., Dai, J.Y., Xiu, Z.L., 2009. Microbial production of 2,3-butanediol from Jerusalem artichoke tubers by *Klebsiella pneumoniae*. Appl. Microbiol. Biotechnol. 82, 847−852.

Syu, M.J., 2001. Biological production of 2,3-butanediol. Appl. Microbiol. Biotechnol. 55, 10−18.

Szambelan, K., Nowak, J., Czarnecki, Z., 2004. Use of *Zymomonas mobilis* and *Saccharomyces cerevisiae* mixed with *Kluyveromyces fragilis* for improved ethanol production from Jerusalem artichoke tubers. Biotechnol. Lett. 26, 845−848.

Thomsen, M.H., Thygesen, A., Thomsen, A.B., 2009. Identification and characterization of fermentation inhibitors formed during hydrothermal treatment and following SSF of wheat straw. Appl. Microbiol. Biotechnol. 83, 447−455.

Ui, S., Takusagawa, Y., Sato, T., Ohtsuki, T., Mimura, A., Ohkuma, M., et al., 2004. Production of L-2,3-butanediol by a new pathway constructed in *Escherichia coli*. Lett. Appl. Microbiol. 39, 533–537.

Wang, A., Wang, Y., Jiang, T., Li, L., Ma, C., Xu, P., 2010. Production of 2,3-butanediol from corncob molasses, a waste by-product in xylitol production. Appl. Microbiol. Biotechnol. 87, 965–970.

Wang, A., Xu, Y., Ma, C., Gao, C., Li, L., Wang, Y., et al., 2012. Efficient 2,3-Butanediol production from cassava powder by a crop-biomass-utilizer, *Enterobacter cloacae* subsp. dissolvens SDM. PLoS One 7, 1–8.

Wilson, C.E., Lucas, H.J., 1936. Stereochemical relationships of the isomeric 2,3-butanediols and related compounds; evidence of Walden inversion. J. Am. Chem. Soc. 58, 2396–2402.

Xian, M., (2015). Sustainable production of bulk chemicals: integration of bio-, chemo-resources and processes.

Xin, F., Basu, A., Weng, M.C., Yang, K.L., He, J., 2016. Production of 2,3-butanediol from sucrose by a *Klebsiella* species. Bioenergy Res 9, 15–22.

Xiu, Z.L., Zeng, A.P., 2008. Present state and perspective of downstream processing of biologically produced 1,3-propanediol and 2,3-butanediol. Appl. Microbiol. Biotechnol. 78, 917–926.

Yang, T., Rao, Z., Zhang, X., Xu, M., Xu, Z., Yang, S.T., 2013. Improved production of 2,3-butanediol in *Bacillus amyloliquefaciens* by over-expression of glyceraldehyde-3-phosphate dehydrogenase and 2,3-butanediol dehydrogenase. PLoS One 8, 1–9.

Yang, T., Rao, Z., Zhang, X., Xu, M., Xu, Z., Yang, S.T., 2017. Metabolic engineering strategies for acetoin and 2,3-butanediol production: advances and prospects. Crit. Rev. Biotechnol. 37, 990–1005.

Zeng, A.P., Biebl, H., Deckwer, W.D., 1990. Effect of pH and acetic acid on growth and 2,3-butanediol production of *Enterobacter aerogenes* in continuous culture. Appl. Microbiol. Biotechnol. 33, 485–489.

Zhang, L., Sun, J., Hao, Y., Zhu, J., Chu, J., Wei, D., et al., 2010. Microbial production of 2,3-butanediol by a surfactant (serrawettin)- deficient mutant of *Serratia marcescens* H30. J. Ind. Microbiol. Biotechnol. 37, 857–862.

CHAPTER 10

Methyl ethyl ketone

Contents

10.1 Introduction

Methyl ethyl ketone, $C_4H_8O_7$, also known as 2-butanone, methyl acetone, ethyl methyl ketone, and MEK, is a ketone with both the methyl group and the adjacent ethyl group. MEK is a colorless liquid with a musty scent that is very similar to acetone. MEK is insoluble in water but is miscible in alcohol, ether, and benzene. It is extremely volatile and flammable by its nature. MEK in the vapor phase is heavier than air, with a density of 0.805 g/cm^3 at 20°C. The registration number of the Chemical Abstracts Service (CAS) for MEK is 78-93-3. MEK has several applications highlighting its use as fuel. The general characteristics of the MEK are shown in Table 10.1 (Gad, 2015).

MEK is commonly used in producing paints, lacquers, varnishes, sticks, resins, gums, nitrocellulose, cells, artificial leather, and other products in industrial applications. It is also used as an aerosol surface cleaner and in the printing industry as well as in the manufacture of dyes and wash insulation agents. The MEK market is projected to rise to USD 3.26 billion and its production to 1754 million tons by 2020 for such chemical versatility (Zhenhua et al., 2006). Paints and coatings have emerged as the leading application segment and a compound annual growth rate of 4.4% from 2016 to 2024 is expected to be registered. Increased expenditure on

Improvements in Bio-Based Building Blocks Production Through Process Intensification and Sustainability Concepts
DOI: https://doi.org/10.1016/B978-0-323-89870-6.00004-4

Table 10.1 General properties of methyl ethyl ketone (MEK) in comparison with other fuels (Hoppe et al., 2016).

Component	Gasoline	Ethanol	2-Methylfuran	MEK
Sum formula	Various	C_2H_6O	C_5H_6O	C_4H_8O
Carbon mass fraction (%)	83.48	52.14	73.15	66.63
Hydrogen mass fraction (%)	13.24	13.13	7.37	11.18
Oxygen mass fraction (%)	3.25	34.73	19.49	22.19
Density (25°C) (kg/m^3)	741	787	907.5	799
Boiling temperature (°C)	35.8–190.4	78	64	80
Vapor pressure (20°C) (kPa)	–	5.8	13.9	10.8
Specific enthalpy of vaporization (kJ/kg air)	–	101.6	35.52	46.1
Stoichiometric air requirement/L (L)	13.97	8.98	10.08	10.52
Lower heating value (MJ/kg)	41.56	26.84	30.37	31.45
Lower heating value (MJ/L)	30.78	21.09	27.63	25.16
Research octane number (*I*)	96.9	109	101.7	117
Motor octane number (*I*)	86.4	89.7	82.4	107
Purity (%)	–	>99	>99	>99

construction activities with rapid growth of the automotive industry in emerging markets in the Asia Pacific and Middle East regions has resulted in high demand for paints and coatings. Paints and coatings include specialty water-borne, powder-borne, and solvent-borne coatings. The high demand for powder coatings in the automotive and electronics industries has significantly increased the product consumption in this application category. Adhesives and printing ink segments are both projected to see rapid growth from 2016 to 2024. Fig. 10.1 shows the growing market for MEK from 2014 to 2024.

Increasing demand from the packaging and publishing sector for printing inks is expected to boost the MEK market in this segment. Due to its characteristics, such as fast drying and better adhesion to substrates such as metals, plastics, and glass, the product is widely used and is a preferred

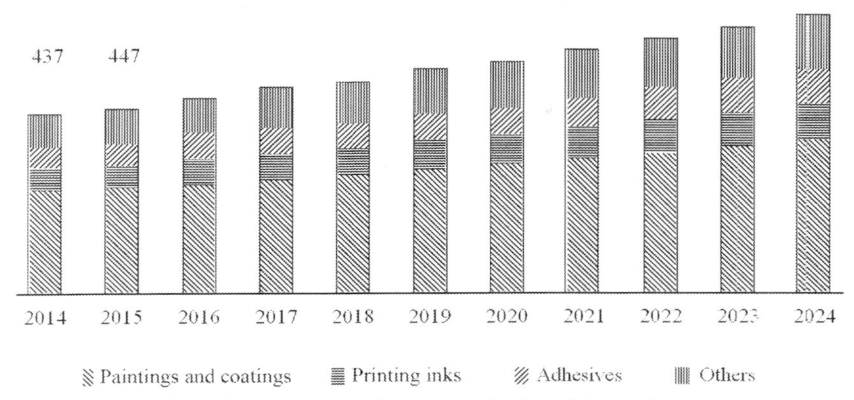

Figure 10.1 Methyl ethyl ketone market by application (kilotons).

solvent in the printing industry. In the next 6 years, advances in end-user industries such as pharmaceuticals, food manufacturing, and cosmetics are expected to drive volumes in the application of printing inks. High-resolution printing is made possible by MEK solvent-based inks used in these industries. Newspapers, stickers, industrial printing, and books are other significant uses for printing inks.

In 2015, Asia Pacific led the global industry and accounted for more than 55% of worldwide production. The high request in emerging nations like China, India, Indonesia, and Thailand is mainly driven by regional volumes of the printing ink market. Due to increased demand for adhesives and printer inks of residential and commercial buildings, the area is expected to continue its dominance over the forecast period. Fig. 10.2 shows the global market of MEK, highlighting the Asia Pacific region as the most predominant in MEK production.

Regulatory bodies have placed strict regulations on MEK producers in relation to the levels of toxicity of the commodity, particularly in developed economies. In the near future, this is expected to curb demand growth. Fluctuating crude oil prices and their downstream derivatives can also affect the profitability of the market directly, as most popular solvents, like MEK, are crude oil downstream derivatives.

The main regulations come for the physicochemical properties of MEK. For example, the vapor of MEK is heavier than air and is soluble in water and some other organic solvents. When they are in concentrations between 1.4% and 11%, they can produce explosive mixtures with oxygen (or air). It can spread over long distances, so it has the potential to

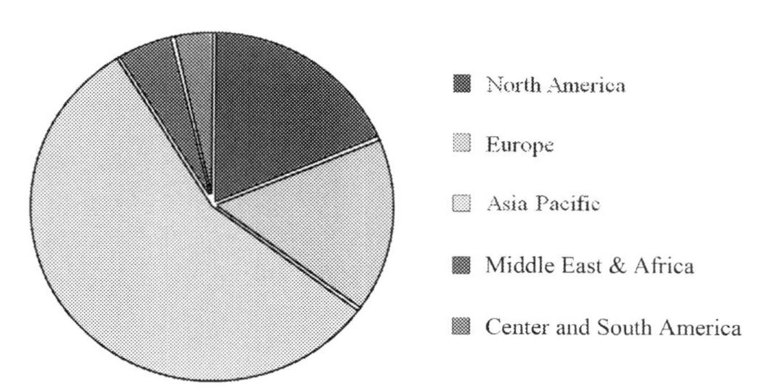

Figure 10.2 Global distribution of the Middle East & Africa market.

ignite from a distance (García, 2008). As far as human harm is concerned, it has been stated that the amount of MEK at which no adverse effects are observed (threshold limit value) is 200 ppm in 8 hours and up to 300 ppm for short exposures of 15 minutes. It was found in a study reported by the Center for Occupational Health and Safety that 143 volunteers reported headache, sore throat, nausea, and general discomfort when exposed to 220 ppm MEK for 4 hours. The neurotoxic effects of MEK, on the other hand, were extensively investigated (Callender, 1995; Mitran et al., 1997). MEK can be produced by two routes, a synthetic route derived from products obtained in the refining of oil, and another route where the presence of microorganisms using renewable raw materials may be involved. Both alternatives will be addressed below.

10.2 MEK production

10.2.1 MEK production from nonrenewable sources

A possible intermediate compound for the processing of hydrocarbons is 1,3-butadiene (1,3-BD). 1,3-BD can be produced by fermenting industrial waste from the steel and biomass industries, such as molasses, cereal mash, starch, wheat, sulfite liquor, and maize starch, using a variety of microorganisms, such as *Aerobacter aerogenes*. However, the most common, and economically feasible way so far is by the petrochemical route.

Usually, steam-cracking is used to produce 1,3-BD, as shown in Fig. 10.3. One of the by-products of the manufacture of ethylene from steam-cracking is BD. At about 850°C, the corresponding feedstocks (C2–C4, naphtha, and oil) are cracked. The steam-cracking furnace produces hydrogen, ethylene, propylene, butadiene, and olefin, and they go

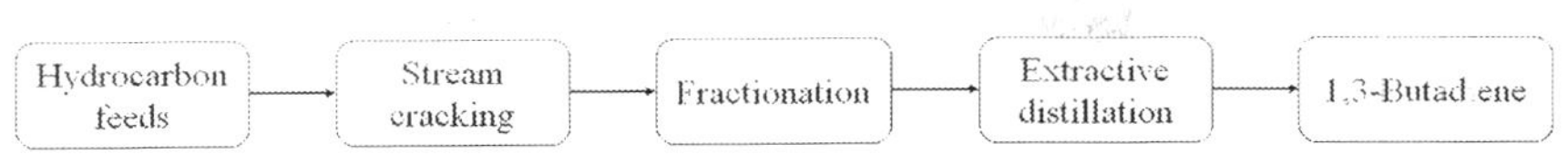

Figure 10.3 Conventional process to manufacture 1,3-BD.

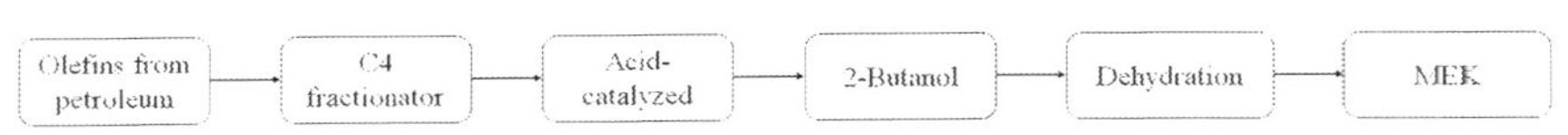

Figure 10.4 Conventional process to manufacture methyl ethyl ketone.

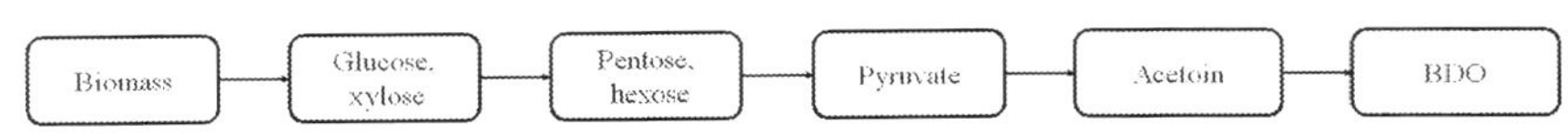

Figure 10.5 Fermentative process to produce 2,3-BD.

through a quencher, a compressor, and a dryer. The hydrocarbons containing more than four carbons (C-5 and higher components) are extracted during this process and the 1,3-BD is purified by extractive distillation. The C4 raffinates and the vinyl acetylene are usually removed by one or two extractive distillations. Thereafter, via a fractionator, the main stream generates the BD to extract the methyl acetylene (White, 2007).

A commercial approach is shown in Fig. 10.4 to produce MEK, which is accomplished by butylene hydration, production of secondary butyl alcohol (SBA) and dehydration of SBA. Butylene is derived from commercial-scale cracking of petroleum. SBA can be obtained from 1-butene and 2-butene hydration based on acid-catalysis. The hydration process is carried out at around 250°C and involves a 75% solution of H_2SO_4; specifically, 1-butene/2-butene is reacted in a reactor and transported to a column of distillation separating the nonreactants from 2-butanol. Furthermore, the SBA is transferred to an intermittent column of distillation. The processed SBA is dehydrated and the MEK is manufactured (Song, 2016), please observe Figs. 10.3–10.5.

As previously mentioned, conventional MEK production involves relatively risky operating conditions. In addition, due to the nature of the process itself, it is clear that there is a dependence on petroleum derivatives. From an environmental point of view, beyond generating a possible product, there is an unhealthy dependence on oil and the consequences of its use.

10.2.2 MEK production from renewable sources

2,3-Butanediol (2,3-BD) is a chemical and fuel building block that can be used in the manufacture of fine chemicals, antifreeze, and the food industry. 2,3-BD is produced through the fermentation of xylose and glucose based on biomass, as shown in Fig. 10.5. The intermediate compounds of 2,3-BD are pentose and hexose; that is, they can be converted into 2,3-BD through glycolysis and reduction. A biochemical route that transforms glucose or xylose into pyruvate is glycolysis.

Adenosine triphosphate and reduced adenine dinucleotide nicotinamide (NADH) are produced through the process. The decrease in acetoin after glycolysis produces what NADHH requires (Celińska and Grajek, 2009). The method of 2,3-BD biofermentation depends on the availability of low-cost biomass or industrial waste gases (Song et al., 2017).

Once 2,3-BD is already obtained, the direct process to MEK is through dehydration. The response pathways of BDO dehydration to an amorphous calcium phosphate (a-CP) catalyst are shown in Fig. 10.6 (Song, 2016). BDO dehydration produces the key BD and MEK products, whereas 2-methylpropanal and 3-buten-2-ol are by-products. A conventional rearrangement, which is the translation of a 1,3-hydride leading to MEK or a 1,3-methyl leading to 1,3-BD, is shown by the first mechanism. A sequential 1,2-elimination of H_2O containing 3-buten-2-ol ($3B_2OL$) is described by the second mechanism. After one more dehydration, BD is created.

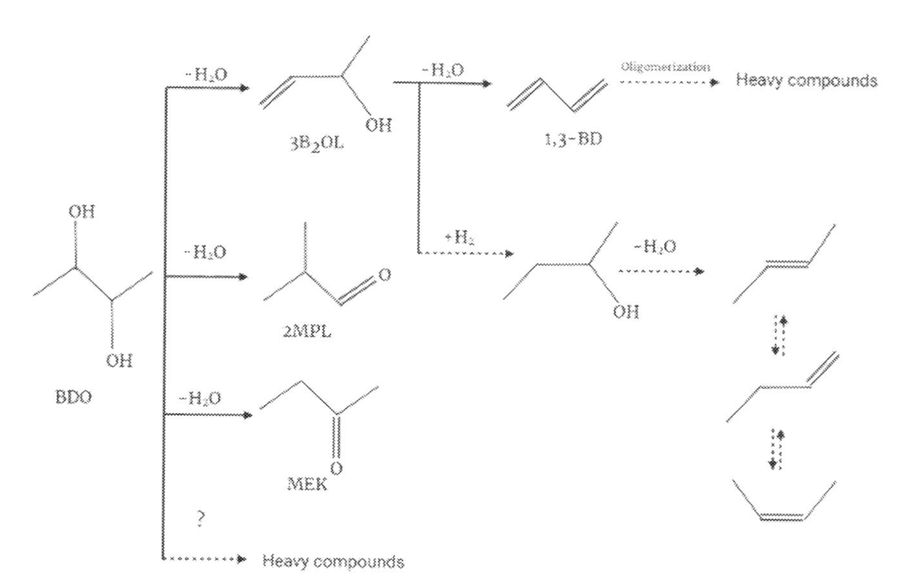

Figure 10.6 2,3-BD dehydration process; solid line for main reaction and dashed lines for minor reactions.

Most of the BDO dehydration-related research studies focus on dehydration catalyst conduction or reaction conditions that generate high yields of the target products, like BD, MEK, and $3B_2OL$, through a special catalyst (Nakamura and Osamura, 1993; Song, 2016; Zhao et al., 2017).

Makshina et al. (2014) have reviewed research on 2,3-BDO dehydration to 1.3-BD and MEK by different catalysts and Duan et al. (2016) have summarized research on 1,3-BD by biuma-derived C4 alcohols, including butanediol dehydration to nonsaturated alcohols. Since the 1940s, the dehydration of 2,3-BDO to 1,3-BD and MEK has been explored. The catalysts used are bentonite clay (Bourns and Nicholls, 1946), oxides of metal and earth (Duan et al., 2015, 2014; Winfield, 1945), zeolites (Bucsi et al., 1994; Lee et al., 2000; Molnár et al., 1988), sulfonic acid group perfluorinated resin(Bucsi et al., 1994), heteropolyacids (Molnár et al., 1988), calcium phosphates (Han et al., 2011; Nikitina and Ivanova, 2016; Tsukamoto et al., 2016), Cs/SiO_2 (Kim et al., 2015), and so on. Among these research groups, a higher selectivity of 1.3-BD ($> 90\%$) was achieved employing SiO_2 supported in CsH_2PO_4 (Tsukamoto et al., 2016). Most of the methods are focused on catalytic dehydration or reaction conditions that generate high yields including 1,3-BD, MEK, and $3B_2OL$.

10.2.2.1 Kinetic equations to methyl ethyl ketone production

An extremely important aspect involved in the intensification of processes for MEK production is the knowledge acquired about the kinetics of MEK production. Based on experimental observations, Fig. 10.6 indicates a possible reaction network using the a-CP catalyst. The reaction network proposed is made based on a theoretical background (Nakamura and Osamura, 1993). 2,3-BD dehydration can be represented as a collection of two parallel routes that produce the main products (1,3-BD, MEK, $3B_2OL$, 2MPL (2-methylpropanal)). The first route is a conventional rearrangement mechanism that occurs through a 1,3-hydride shift leading to MEK or a 1,2-methyl shift leading to 2MPN. A sequential 1,2-removal of water leading to the formation of $3B_2OL$ after the first dehydration and 1,3-BD after the second (Kim et al., 2016) is the second direction. Impurities in the reaction products of this analysis is not included, as the total amount of minor butene isomers and heavy compounds considered as impurities for all experiments is less than 0.3 wt.%, and the kinetic model is greatly simplified by disregarding impurities.

The following reactions identify the main pathway of 2,3-BDO dehydration:

$$\underset{(2,3-BDO)}{C_4H_{10}O_2} \xrightarrow{r_1} \underset{(3B_2OL)}{C_4H_8O} + H_2 \tag{10.1}$$

$$\underset{(3B_2OL)}{C_4H_8O} \xrightarrow{r_2} \underset{(1,3-BD)}{C_4H_6} + H_2O \tag{10.2}$$

$$\underset{(2,3-BDO)}{C_4H_{10}O_2} \xrightarrow{r_3} \underset{(MEK)}{C_4H_8O} + H_2O \tag{10.3}$$

$$\underset{(2,3-BDO)}{C_4H_{10}O_2} \xrightarrow{r_4} \underset{(2MPL)}{C_4H_8O} + H_2O \tag{10.4}$$

As a power-law reaction, the kinetic reaction is reduced as follows:

$$k_i = k_{T_{ref},i} \exp\left(\frac{-E_i}{R}\left(\frac{1}{T} - \frac{1}{T_{ref}}\right)\right) \tag{10.5}$$

Table 10.2 displays the approximate kinect power-law parameters.

This kinetic data set has been used as a basis for the kinetic understanding of MEK production from 2,3-BD for both conventional and intensified production schemes. Both cases will be addressed in the following lines.

10.2.3 Production ok methyl ethyl ketone through process intensified schemes

MEK may be produced by direct fermentation, but with a yield of approximately 0.004 g $MEK/g_{glucose}$, its performance is poor (5). A

Table 10.2 Estimated kinetic parameters of the power-law model (Song, 2016).

Model parameter	Value
E_1 (J/mol)	2.33E + 05
E_2 (J/mol)	2.82E + 05
E_3 (J/mol)	1.93E + 05
E_4 (J/mol)	1.66E + 05
$K_{T_{ref},1}$ $(mol^{(1-n_1)}\ m^{3(n_1-1)}/s)$	7.45E − 04
$k_{T_{ref},2}$ $(mol^{(1-n_1)}\ m^{3(n_1-1)}/s)$	4.41E − 04
$k_{T_{ref},3}$ $(mol^{(1-n_1)}\ m^{3(n_1-1)}/s)$	6.64E − 04
$k_{T_{ref},4}$ $(mol^{(1-n_1)}\ m^{3(n_1-1)}/s)$	1.27E − 04
n_1, n_3, n_4	1.87E − 02
n_2	1.46E − 01

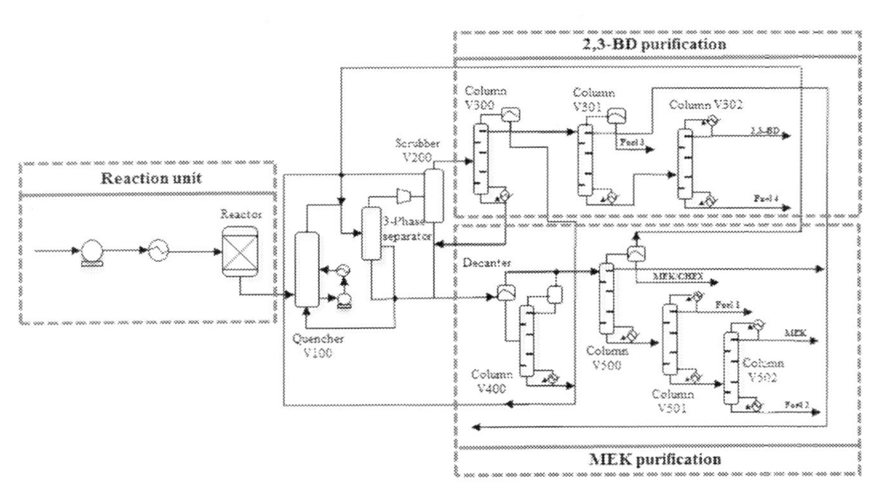

Figure 10.7 Production of methyl ethyl ketone in a two-step process by Song et al. (2017).

promising alternative to MEK is intermediate with 2,3-BD. Interestingly, the output of 2,3-BD by fermentation close to the theoretical limit of 0.5 g 2,3-BD/$g_{glucose}$ is relatively high on this road (Bourns and Nicholls, 1946; Duan et al., 2016).

In an interesting review, Song et al. (Molnár et al., 1988) proposed an alternative to the development and purification of MEK from 2,3-BD. In their plan, they proposed a reactor-based development scheme followed by a sequence of separation columns and decanters, for a total of 10 separation units and a reactor (see Fig. 10.7). Distillation columns are the bulk of all separation devices.

The general characteristics of the scheme in Fig. 10.7, reported by Song et al. (2017) are shown in Table 10.3.

Process Intensification (PI) is an alternative to improving a system. PI is defined by five characteristics: reduced equipment size, improved process performance, reduced stock of equipment, decreased usage of utilities and raw materials and increased process equipment efficiency (Bucsi et al., 1994). In several cases where chemical reactions include, for example, reactive distillation (RD), in which even the reactor functions as a separation mechanism, this technique has been successfully proven, so the intensification of the processes enables conversion and phase balance limitations to be removed.

Table 10.3 General parameters of process route of Fig. 10.7.

Description	V100	V200	V300	V301	V302	V400	V500	V501	V502
Number of stages	5	5	20	80	56	10	30	62	16
Pressure (kg/cm^2 g)	1.4	3.9	3.5	3.8	3.5	1.5	1.1	−0.35	−0.3
Feed temperature (°C)	180	90	38.5	38	44.4	42.2	56.9	106	72.2
Overhead temperature (°C)	51	38	40.6	42.1	40.8	94.1	85	52.8	69
Bottom temperature (°C)	80	128	138	44.4	44.1	131.1	106	72.2	175
Heat duty (MMKcal/h)	–	–	0.76	2.16	0.93	1.72	2.25	2.67	0.65

The changes documented by the use of RD are reasonably diverse and in different case studies. Souza et al. (2017) have recorded, for example, a substantial improvement in the purity obtained by the production of triacetin compared to the purity obtained by the traditional method. Pöpken et al. (2001), on the other hand, reported a decrease in energy requirements, increased selectivity, and conversion when processing methyl acetate.

In certain cases where the use of RD has been reported, capital cost and energy cost savings of 20% have been reported (Lutze et al., 2010). Biofuel production has not been excluded from the use of RD but its use is stated mainly for the production of biodiesel. Kiss et al. (2008), for example, suggested processing biodiesel using multiple heterogeneous catalysts (niobic acid, sulfated zirconia, sulfated titania, and sulfate tin oxide). The response time was improved by the use of the RD, efficiency was increased, and the size of the equipment was decreased, achieving lower capital costs. Similarly, other authors reported improvements in productivity and cost of production for the production of biodiesel (Pérez-Cisneros et al., 2016; Poddar et al., 2017).

In systems with some chemical and phase conditions coexist (Luyben and Yu, 2009), RD is highly desirable. According to Shah et al. (2012), it is important to evaluate the feasibility of such a method before a formal proposal for RD. In this case, the analysis should be accompanied by the recommendations that all these criteria should be fulfilled by pointing out (1) existence of more than one element, (2) a correlation exist between the reaction and the temperature separation, (3) operating pressure and temperature are not near the critical area of the components in question, and (4) volatility of the components.

Taking as base a feed stream of 1000 kg/hour of 2,3-BD Torres-Vinces et al. (2020) presented an intensified alternative to produced MEK based on the previous work presented by Song et al. (2017) (see Fig. 10.8).

In the reference case of Fig. 10.7, the number of decanters considered by Song et al. (2017), which results in several waste streams, is fair. The explanation for this design is that the system is thermodynamically complex to purify. There are four azeotropic components: three heterogeneous between 1,3-BD and H_2O, respectively; and MEK-H_2O and 2MPL$-H_2O$, and one azeotropic homogeneous between $3B_2OL$-and H_2O, respectively (see Fig. 10.8). Conventional columns are not always an ideal choice for azeotropic separations because there is considerable energy demand for this process.

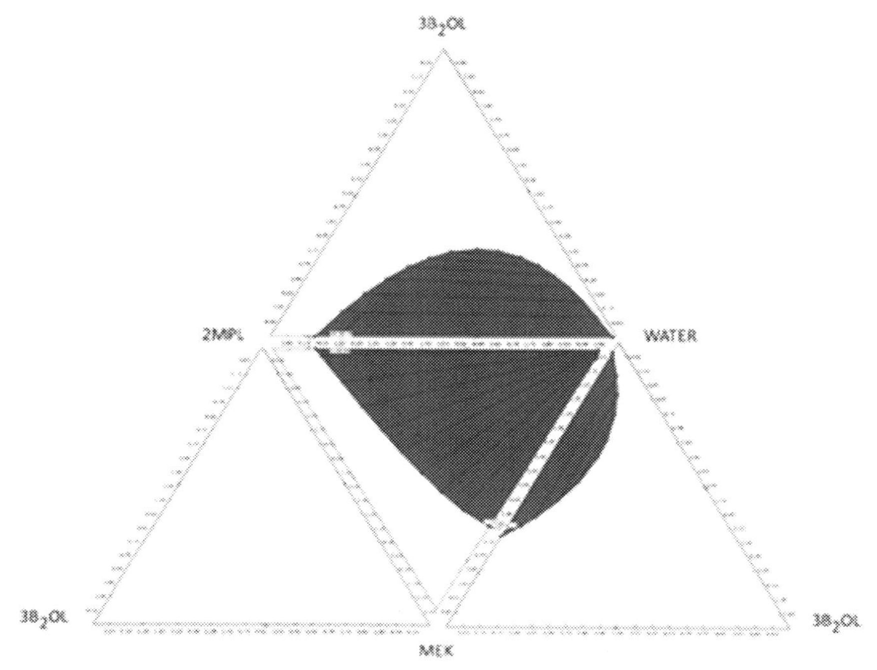

Figure 10.8 Quaternary maps (mole basis) of the components coming from RD.

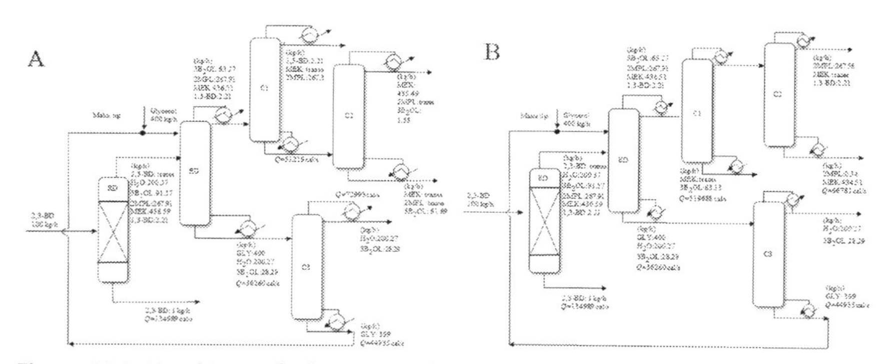

Figure 10.9 Novel intensified proposals for methyl ethyl ketone production.

As product of their research, two intensified alternatives were presented. The base of both intensified alternatives was a RD column and an extractive distillation column. Glycerol was the best choice for being used as a solvent in the extractive distillation column after an exhaustive analysis and multiple tests of the simulator (Fig. 10.9).

Both enhanced alternatives achieve a recovery constraint of at least 98%wt. for all components and a purity constraint of 99.5%wt. for MEK, 99%wt. for 3-buten-2-ol, 99%wt. for 2MPL, 99%wt. for 1,3-BD and 99.99%wt. for glycerol, since 99.5%wt. is the lowest purity to be considered as fuel (26.2% for MEK). Topologically, the only distinction between the two proposals is that purification is carried out in scheme A in a direct sequence following the extractive distillation column, compared to scheme B. The first efficiency index was the energy usage to compare all alternatives. The main aim of these processes is to realize the amount of energy expended in units like MJ/kg_{MEK}. Moreover, as MEK will be used for spark-ignition motors, the amount of energy expended and the amount of energy produced for MEK ignition would be an interesting comparison. The eco-indicator 99 (EI99) and CO_2 emissions were the second and third efficiency index.

As result, Tables 10.4 and 10.5 show the main design parameters to design those two intensified alternatives.

Table 10.4 Design parameters for intensified alternative A.

	RD	ED	C1	C2	C3
Number of stages	80	50	80	46	30
Reactive stages	10–80	–	–	–	–
Solvent (kg/h)	–	400	–	–	–
Reflux ratio	2.95	0.5	5.488	4.712	0.039
Feed stage	76	5, 46	23	25	14
Operative pressure (kPa)	101.353	101.353	101.353	101.353	101.353
Distillate flowrate (kg/h)	999	770	270	437.6	230
Reboiler duty (kcal/h)	485,962	130,535	184,390	256,145	1,617,65
CO_2 emissions (ton/h)	7.07				
Eco-Ind (points/ year)			1.825E + 06		

ED - Extractive Distillation.

Table 10.5 Design parameters for intensified alternative B.

	RD	EC	C1	C2	C3
Number of stages	80	50	56	50	30
Reactive stages	10–80	–	–	–	–
Solvent (kg/h)	–	400	–	–	–
Reflux ratio	2.95	0.5	14.1	7.01	0.039
Feed stage	76	5, 46	5	5	23
Operative pressure (kPa)	101.353	101.353	101.353	101.353	101.353
Distillate flowrate (kg/h)	999	770	706.65	270.65	230
Reboiler duty (kcal/h)	485,962	130,535	1,130,280	229,264	161,765
CO_2 emissions (ton/h)	13.42				
Eco-Ind (points/year)			4.75E + 06		

The RD column registered a 99.86% conversion and 44% selectivity for MEK. An extractive distillation column and three conventional columns were used to purify the effluent coming from the reactor (in both direct and indirect arrangement after extractive column). The most promising energy requirement was 2790 kcal/kg_{MEK} (11.6 MJ/kg_{MEK}) after evaluation of the direct scheme (scheme A). Concerning the environmental effect of greenhouse gas emissions, 7.07 ton CO_2/hour was recorded by scheme A.

Although this process can be considered as an intensified scheme for MEK production, it is also true that it can be considered as an intensified scheme for MEK production and separation. Note that in the production process there is a RD column; additionally, intensified technology is used for the separation of the effluent that comes from it, an extractive distillation column.

In the following section, separation schemes that have been reported to purify the effluent from the 2,3-BD dehydration process will be discussed.

10.3 Purification of MEK through intensified process

Despite reasonably high yields for 2,3-BD fermentation and further dehydration, the downstream process has not yet been well explored. In addition, by-products such as isobutyraldehyde (2MPL) and 2,3-BD are valuable products that may increase the value of the bio-based refinery for the production of MEK, both of which have significant participation in the food industry and biosynthesis of isobutanol (Xiao and Lu, 2014). 2,3-BD has a variety of implementations. For example, it can be used as an intermediate for the production of rubber. It can be used as an antifreeze compound due to its low freezing point.

In the manufacture of fumigants, perfumes and inks, explosives and suppressants, 2,3-BD has also demonstrated its potential (Celińska and Grajek, 2009). In the other hand, isobutyraldehyde (2MPL) is used as an additive for the preparation of fragrances and flavors. It is also used as a precursor for the manufacture of plasticizers, isobutyric acid, and isobutanol (Rodriguez and Atsumi, 2012). Notice, despite the promising thermodynamic properties described earlier, MEK purification is still challenging as two azeotropes are present in the MEK/2MPL/2,3-BD/water mixture (See Fig. 10.8).

The proposals which properly analyze a MEK/IBA/2,3-BD/water mixture are very scarce, because of the oil-based methodology used to generate the MEK. The purification method does not cover the mixture purified in this work although some approaches are available to the MEK purification. In addition, they are mixtures which are not thermodynamically complex in the quaternary mixture. For example, Smetana et al. (1996) have approached MEK—water purification based on a membrane method, on the other hand, Lloyd (1984) has patented a purification method of the MEK—ethyl acetate blend the same mixture that Murphy (2000) investigated.

The conceptual design of four alternatives for purifying of the MEK/2MPL/2,3-BD/water mix was recently proposed by Penner et al. (2017) but it is based on the entire proposal on distillation columns and decanters. It does not, however, suggest a rigorous design strategy with consistent target functions and constraints on recovery and purity. As always, the first option for this difficult separation is distillation (Fig. 10.10).

In a recent review, Sánchez-Ramírez et al. (2021) proposed an improved approach to MEK purification. The ratio of all parts is 65% MEK, 18, 10, 3, and 7%wt. (see Table 10.6), a relatively typical dehydration reactor outlet (Tran and Chambers, 1987; Zhao et al., 2016).

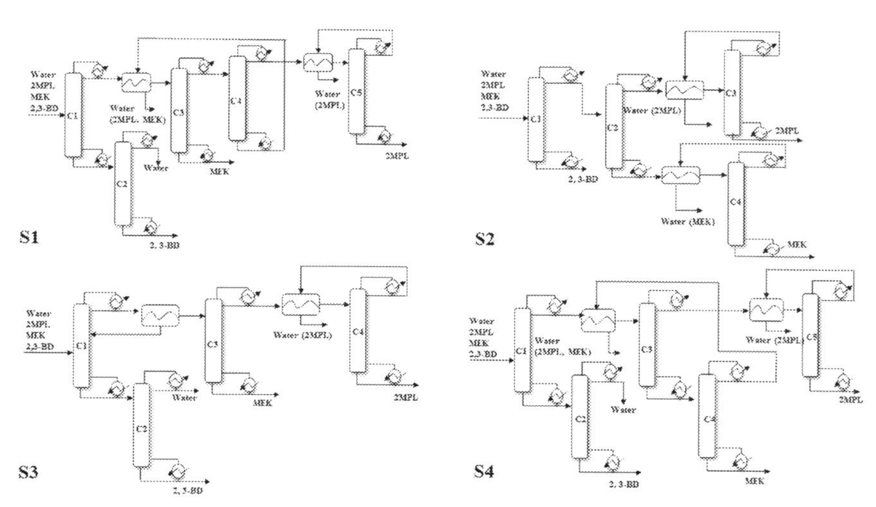

Figure 10.10 Pure distillation alternatives for MEK purification. (A) scheme S1, (B) scheme S2, (C) scheme S3, (D) scheme S4.

Table 10.6 Feed characterization.

Feed concentration (wt.%)				Vapor fraction	Flowrate (kg/h)	Temperature (K)
Water	2-MPL	2,3-BD	MEK			
0.18	0.07	0.1	0.65	0	11764.7	298

The scheme presented by Sánchez-Ramírez et al. (2021) is an intensified hybrid scheme initially composed of a liquid–liquid extraction (LLE) column, considering p-xylene as solvent. The use of this LLE column helps to separate more efficiently the azeotropes formed by the various compounds in the feed stream. Hybrid systems are a good example of intensified processes. Hybrid processes focused on both LLE and distillation columns have proved their effectiveness in reducing global downstream energy consumption. The key explanation for the reduction of energy requirements lies in the fact that the LLE column helps to break down the thermodynamic interactions between the components and thus decreases the energy requirement (Fig. 10.11).

To evaluate the improvement that can be obtained with the intensified scheme, the alternatives in Figs. 10.9 and 10.10 were evaluated with various performance indices. In a multi-objective optimization framework, four objective functions were considered jointly and in early design stages:

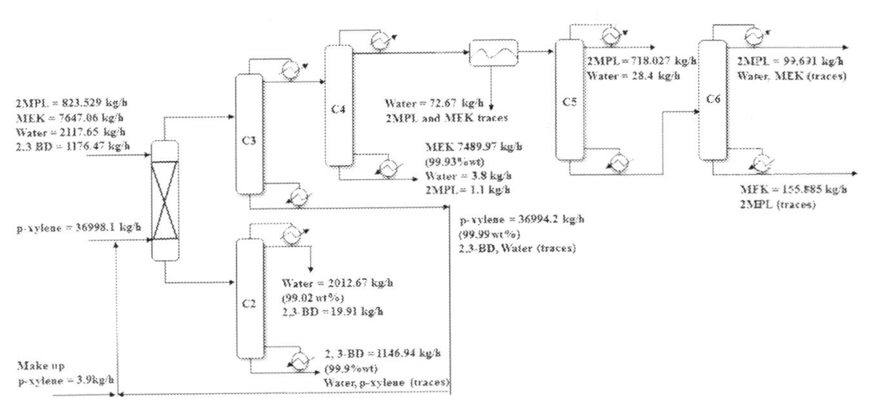

Figure 10.11 Intensified alternatives for the MEK purification.

total annual cost (TAC), EI99, inherent process safety (IR), and condition number as an indicator of process controllability. The main goal was to minimize all four objective functions. To optimize the case of study, this work employs a hybrid stochastic optimization algorithm, Differential Evolution with Tabu List.

As initial results, Sánchez-Ramírez et al. (2021) showed the null feasibility of most of the schemes presented in Fig. 10.10. That is, of the four schemes analyzed in Fig. 10.10, only scheme S2 is feasible to generate designs with high purities and high recoveries. Schemes seem close at first sight; however, the location of decanters is a crucial difference. The function of the decanter is to facilitate separation when there are three phases. However, because of the ratio between parts, the decanter is not always capable of completely separating these three phases. Observe Fig. 10.12: before entering the decanter, we take the mixture as examples to purify MEK. Please note that the mixture of scheme S2 is the nearest to the binary mixture MEK—water in the quaternary diagram (binary ax MEK—water).

That is, the column C1 plays an important part in the separation of the mix. The schemes S1, S3, and S4 aim to extract as much water as possible from the azeotropic zone to the bottom of the column, while only 2,3-BD is sent from the S2 to the bottom of the column. Fig. 10.12 shows the difference and why high recoveries in schemes S1, S3, and S4 cannot be obtained, and Fig. 10.11 shows a complete mass balance for the most promissory scheme.

Once the low feasibility of the conventional schemes was analyzed, by means of the Pareto fronts analysis, it was possible to determine the design

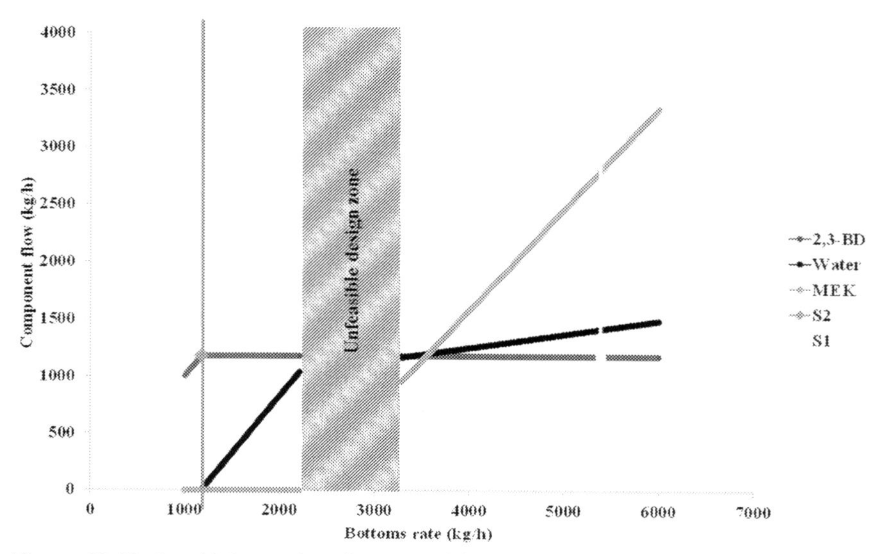

Figure 10.12 Sensitivity analysis for C1 and first decanter in schemes S1, S3, and S4.

parameters that allow minimizing the objective functions. The main design parameters as well as the objective functions are shown in Tables 10.7 and 10.8.

All separation systems have shown very interesting results following the optimization process. The most promising solution, an LLE-based hybrid process, was the only way that the whole food mix was regenerated and purified. The hybrid method increases energy use, energy advantage and thermodynamic performance in contrast to the distillation scheme, unlike pure and not energy-sustainable distillation systems. Furthermore, enormous energy savings were seen, which are therefore observed in TAC, EI99, and IR efficiency parameters. In view of the qualitative nature of the condition number, the percentage of changes to the other systems in relation to the intensified system can hardly be defined. But it can be seen that there is enough incentive to design a stable control mechanism for the improved system, with good economic and environmental performance. In addition, the current connection between many design alternatives to the objective functions, demonstrating the reflux ratio and heat duties, may be known. This meant that the existing economic, environmental, controllability, and security needs could be met through the tendencies and opportunities in zones to formulate separation alternatives.

Table 10.7 Design parameters and performance indexes for the intensified scheme.

	LLE	C2	C3	C4	C5
Number of stages	10	33	45	45	54
Reflux ratio		3.483	0.529	16.636	5.01
Feed stage	1, 10	4	27	5	23
Column diameter (m)	1.455	1.285	1.407	1.544	1.098
Operative pressure (kPa)	101.353	101.353	101.353	101.353	101.353
Distillate flowrate (kmol/h)		111.997	123.297	19.193	12.9292
Condenser duty (kW)		5776	1693	3191	694
Reboiler duty (kW)		6354	4125	3202	727
η (%)	32.56				
TAC (USD/year)	7,903,251				
Eco-Ind (points/year)	1,338,593				
Condition number	88,121				
IR (probability/year)	0.0014087				

Table 10.8 Objective function for pure distillation schemes.

	S1	S2	S3	S4	Hybrid-intensified
TAC (USD/year)	104,719,750	31,011,553	153,136,510	4,435,273	7,903,251
Eco-Ind (points/year)	2,993,581,413	14,669,116	16,200,579	891,801,275	1,338,593
Condition number	3.8	3.99	4.78	147.5	88,121
IR (probability/year)	0.00167156	0.0013323	0.00133414	0.00166587	0.0014087

10.4 Conclusion and future insights

Throughout this chapter, several intensified alternatives for the production of MEK were shown. Given the nature of the MEK production process starting from 2,3-BD, several PI techniques can be applied in both the reaction and separation zones. In the reaction section, through the use of a RD column, a notable improvement in energy consumption was generated compared to the conventional production alternative. On the other hand, approximating only the separation section, using an intensified hybrid scheme, improvements were observed in several indicators. The intensified scheme showed a significant improvement in TAC, EI99, inherent safety, and condition number, indicators of cost, environmental impact, safety, and controllability, respectively.

Currently, the production process for the generation of MEK from petroleum derivatives is more widely used. However, in order to generate a production process from renewable raw material, the critical point is the production of 2,3-BD. In other words, the limiting factor for MEK production from biomass is the fermentation route that generates 2,3-BD. Once this fermentation process increases its productivity, it is clear that MEK generation will improve substantially.

On the other hand, another aspect to improve is the catalytic process involved in the dehydrogenation process of 2,3-BD to MEK is also important. Although the use of the RD column generates a substantial improvement in terms of yields, research in different catalytic processes can generate higher yields and conversions to MEK.

References

Bourns, A.N., Nicholls, R.V.V., 1946. The catalytic action of aluminium silicates I. The dehydration of butanediol-23 and butanone-2 over activated morden bentonite. Can. J. Res. 25, 80–88.

Bucsi, I., Molnár, A., Barók, M., 1994. Transformation of 1,2-diols over perfluorinated resinsulfonic acids (Nafion-H). Tetrahedron 50, 8195–8202.

Callender, T.J., 1995. Neurotoxic impairment in a case of methylethyl–ketone exposure. Arch. Environ. Health 50, 392.

Celińska, E., Grajek, W., 2009. Biotechnological production of 2,3-butanediol – current state and prospects. Biotechnol. Adv. 27, 715–725.

Duan, H., Sun, D., Yamada, Y., Sato, S., 2014. Dehydration of 2,3-butanediol into 3-buten-2-ol catalyzed by ZrO_2. Catal. Commun. 48, 1–4.

Duan, H., Yamada, Y., Sato, S., 2015. Efficient production of 1,3-butadiene in the catalytic dehydration of 2,3-butanediol. Appl. Catal. A Gen. 491, 163–169.

Duan, H., Yamada, Y., Sato, S., 2016. Future prospect of the production of 1,3-butadiene from butanediols. Chem. Lett. 45, 1036–1047.

Gad, S.E., 2015. Methyl ethyl ketone. J. Chem. Heal. Saf. 34–36.

García, H.D., 2008. Spacecraft Water Exposure Guidelines for Selected Contaminants, vol 3. National Academy of Sciences Press.

Han, Y., Kim, H., Hong, J., Jho, H., 2011. Fabrication method of 1,3-butadiene and 2-butanone from 2,3-butanediol. Abstract.

Hoppe, F., Heuser, B., Thewes, M., Kremer, F., Pischinger, S., Dahmen, M., et al., 2016. Tailor-made fuels for future engine concepts. Int. J. Engine Res. 17, 16–27.

Kim, T.Y., Baek, J., Song, C.K., Yun, Y.S., Park, D.S., Kim, W., et al., 2015. Gas-phase dehydration of vicinal diols to epoxides: dehydrative epoxidation over a Cs/SiO_2 catalyst. J. Catal. 323, 85–99.

Kim, W., Shin, W., Lee, K.J., Song, H., Kim, H.S., Seung, D., et al., 2016. 2,3-Butanediol dehydration catalyzed by silica-supported sodium phosphates. Appl. Catal. A Gen. 511, 156–167.

Kiss, A.A., Dimian, A.C., Rothenberg, G., 2008. Biodiesel by catalytic reactive distillation powered by metal oxides. Energy Fuels 22, 598–604.

Lee, J., Grutzner, J.B., Walters, W.E., Delgass, W.N., 2000. The conversion of 2,3-butanediol to methyl ethyl ketone over zeolites. Stud. Surfaces Sci. Catal 130, 2603–2608.

Lloyd, B., (1984). Separation of ethyl acetate from methyl ethyl ketone by extractive distillation. 4470881.

Lutze, P., Gani, R., Woodley, J.M., 2010. Process intensification: a perspective on process synthesis. Chem. Eng. Process. Process Intensif. 49, 547–558.

Luyben, W.L., Yu, C.-C., 2009. Reactive Distillation Design and Control. John Wiley & Sons, Inc., Hoboken, NJ, USA.

Makshina, E.V., Dusselier, M., Janssens, W., Degrève, J., Jacobs, P.A., Sels, B.F., 2014. Review of old chemistry and new catalytic advances in the on-purpose synthesis of butadiene. Chem. Soc. Rev. 43, 7917–7953.

Mitran, E., Callender, T., Orha, B., Dragnea, P., Botezatu, G., 1997. Neurotoxicity associated with occupational exposure to acetone, methyl ethyl ketone, and cyclohexanone. Environ. Res. 73, 181–188.

Molnár, Á., Bucsi, I., Bartók, M., 1988. Pinacol rearrangement on zeolites. Stud. Surf. Sci. Catal. 41, 203–210.

Murphy, C.D., (2000). Process of recovering methyl ethyl ketone from an aqueous mixture of methyl ethyl ketone and ethanol. 6121497.

Nakamura, K., Osamura, Y., 1993. Theoretical study of the reaction mechanism and migratory aptitude of the pinacol rearrangement. J. Am. Chem. Soc. 115, 9112–9120.

Nikitina, M.A., Ivanova, I.I., 2016. Conversion of 2, 3-butanediol over phosphate catalysts. Chem. Cat. Chem. 8, 1346–1353.

Penner, D., Redepenning, C., Mitsos, A., Viell, J., 2017. Conceptual design of methyl ethyl ketone production via 2,3-butanediol for fuels and chemicals. Ind. Eng. Chem. Res. 56, 3947–3957.

Pérez-Cisneros, E.S., Mena-Espino, X., Rodríguez-López, V., Sales-Cruz, M., Viveros-García, T., Lobo-Oehmichen, R., 2016. An integrated reactive distillation process for biodiesel production. Comput. Chem. Eng. 91, 233–246.

Poddar, T., Jagannath, A., Almansoori, A., 2017. Use of reactive distillation in biodiesel production: a simulation-based comparison of energy requirements and profitability indicators. Appl. Energy 185, 985–997.

Pöpken, T., Steinigeweg, S., & Gmehling, J. 2001. Synthesis and hydrolysis of methyl acetate by reactive distillation using structured catalytic packings: experiments and simulation. Ind. Eng. Chem. Res. 40(6), 1566–1574.

Rodriguez, G.M., Atsumi, S., 2012. Isobutyraldehyde production from *Escherichia coli* by removing aldehyde reductase activity. Microb. Cell Fact. 11, 1−11.
Sánchez-Ramírez, E., Quiroz-Ramírez, J.J., Contreras-Zarazua, G., Hernández-Castro, S., Alcocer-García, H., Segovia-Hernández, J.G., (2021). Intensified alternative to purify methyl-ethyl ketone in a framework of green process. Energy 220.
Shah, M., Kiss, A.A., Zondervan, E., De Haan, A.B., 2012. Chemical engineering and processing : process intensification a systematic framework for the feasibility and technical evaluation of reactive distillation processes. Chem. Eng. Process. Process Intensif. 60, 55−64.
Smetana, J.F., J.L. Falconer, R.D. Noble, (1996). Separation of methyl ethyl ketone from water by pervaporation using a silicalite membrane. J. Membr. Sci. 127−130.
Song, D., 2016. Kinetic model development for dehydration of 2,3-butanediol to 1,3-butadiene and methyl ethyl ketone over an amorphous calcium phosphate catalyst. Ind. Eng. Chem. Res. 55, 11664−11671.
Song, D., Yoon, Y., Lee, C., 2017. Chemical engineering research and design conceptual design for the recovery of 1,3-butadiene and methyl ethyl ketone via a 2,3-butanediol-dehydration process. Chem. Eng. Res. Des. 123, 268−276.
Souza, T.F.C., Ferreira, N.L., Marin, M., Guardani, R., 2017. Glycerol esterification with acetic acid by reactive distillation using hexane as an entrainer. Int. J. Chem. Eng. Appl. 8, 344−350.
Torres-Vinces, L., Contreras-Zarazua, G., Huerta-Rosas, B., Sánchez-Ramírez, E., Segovia-Hernández, J.G., 2020. Methyl ethyl ketone production through an intensified process. Chem. Eng. Technol. 43, 1433−1441.
Tran, A.V., Chambers, R.P., 1987. The dehydration of fermentative 2,3-butanediol into methyl ethyl ketone preparation of solid acid catalysts properties of catalyst. Biotechnology 29, 343−351.
Tsukamoto, D., Sakami, S., Ito, M., Yamada, K., Yonehara, T., 2016. Production of bio-based 1,3-butadiene by highly selective dehydration of 2,3-butanediol over SiO_2-supported cesium dihydrogen phosphate catalyst. Chem. Lett. 45, 831−833.
White, W.C., 2007. Butadiene production process overview. Chem. Biol. Interact. 166, 10−14.
Winfield, M.E., 1945. The catalytic dehydration of 2,3-butanediol to 1,3-butadiene. J. Counc. Sci. Ind. Res. 18, 412−423.
Xiao, Z., Lu, J.R., 2014. Strategies for enhancing fermentative production of acetoin: a review. Biotechnol. Adv. 32, 492−503.
Zhao, J., Yu, D., Zhang, W., Hu, Y., Jiang, T., Fu, J., et al., 2016. Catalytic dehydration of 2,3-butanediol over P/HZSM-5: effect of catalyst, reaction temperature and reactant configuration on rearrangement products. RSC Adv 6, 16988−16995.
Zhao, L., Lyu, X., Wang, W., Shan, J., Qiu, T., 2017. Comparison of heterogeneous azeotropic distillation and extractive distillation methods for ternary azeotrope ethanol/toluene/water separation. Comput. Chem. Eng. 100, 27−37.
Zhenhua, L., Wenzhou, H., Hao, M., Kai, Q., 2006. Development and commercial application of methyl-ethyl-ketone production technology. Chin. J. Chem. Eng. 14, 676−684.

CHAPTER 11

Lactic acid

Contents

11.1 Lactic acid

Lactic acid ($C_3H_6O_3$) is an organic acid that can be produced by chemical synthesis or fermentation from different sources of carbohydrates such as glucose (starch), maltose (produced by specific enzymatic conversion of starch), sucrose (from syrups, juices, and molasses), lactose (serum), among others (Keller, 2007). The lactic acid that is produced commercially today, comes mainly from glucose fermentation. Its molecule contains an asymmetrical carbon atom. Lactic acid exhibits optical isomerism: L (+) or D (−), although, it is more common to find it as a racemic mixture. Unlike the isomer D (−), the configuration L (+) is metabolized by the human organism (Altıok, 2004). The two optically active forms such as the racemic form are in a liquid state, being colorless and soluble in water. In their purest form, they are highly hygroscopic low melting point solids, which

Improvements in Bio-Based Building Blocks Production Through Process Intensification and Sustainability Concepts
DOI: https://doi.org/10.1016/B978-0-323-89870-6.00013-5

Table 11.1 Some physicochemical properties of lactic acid (Serna-Cock and Stouvenel, 2005).

Formula	$C_3H_6O_3$
Molecular weight	90.08
Melting point L (+) and D (−)	52.8°C−54°C
DL (according to its composition)	16.8°C−33°C
Boiling point	125°C−140°C
pK_a	3.87 a 25°C

is difficult to establish due to the extreme difficulty of producing acid anhydrously. It is for this reason that a range of 18°C − 33°C is used. The boiling point of the anhydrous product is between 125°C and 140°C. The two isomeric forms of lactic acid can be polymerized and polymers with different properties can be produced depending on the composition. The physicochemical properties of lactic acid are shown in Table 11.1 (Serna-Cock and Stouvenel, 2005).

11.1.1 Uses of lactic acid

Lactic acid has various applications in different types of industries, such as: food, chemical, medical, pharmaceutical, and cosmetic, including biodegradable polymers (Vick Roy, 1985; Suriderp, 1995; Chang et al., 1999; Danner et al., 2002; Park et al., 2003).

1. Food industry: as an acidulant and preservative.
2. Chemical industry: as a solubilizer and as a pH controller agent.
3. Tanning industry: to soak leather and refloatation.
4. Paint and resin industry: as a biodegradable solvent.
5. Pharmaceutical industry: production of medicines for their iron and calcium salts.
6. Textile industry: assists in dyeing and printing.
7. Agriculture: as an acidulant.
8. Plastics industry: precursor to polylactic acid (PLA), a biodegradable polymer with interesting uses in industry and medicine; the latter is considered the main application of acid and the cause for which its demand has increased considerably.

Some of the most important chemicals for each of these industries are (Corma et al., 2007):

1. Lactide, for the development of PLA or polylactide and other copolymers.

2. Lactates are used as excellent solvents that could replace toxic and halogenated solvents.
3. Lactate esters are also used as plasticizers in cellulose and vinyl resins and improve the detergent properties of ion surfactants.
4. 1,2-Propanediol is used as solvent to produce unsaturated polyester resins, drugs, cosmetics, and foods, it is also used as a thaw and antifreeze liquid.
5. Acrylic acid is used for building blocks, as well as surface coatings, textiles, adhesives, paper treatment, leather, fibers, detergents, etc.
6. The 2,3-Pentanedione has applications such as flavoring agent, photo initiator and biodegradable solvent and is a useful intermediate for the synthesis of a variety of products.
7. Pyruvic acid is used as precursors in the synthesis of drugs and agrochemicals.

Depending on its use, lactic acid quality specifications are determined. The purity required for the application of lactic acid is for the food industry 80% purity, pharmaceutical 88% and industry 80%–90% (Long and Lee, 2017).

11.1.2 Market and demand for lactic acid

In recent years the global lactic acid market has witnessed high growth, due to the growing demand for its applications (biodegradable polymers, food and beverages, pharmaceuticals, personal care products and others). Fig. 11.1 global demand for lactic acid over the past decade, which is estimated to be between 200,000 and 350,000 tons/year, with annual growth rates between 12% and 15% (Abdel-Rahman et al., 2013).

Although the lactic acid molecule is found naturally in plants, microorganisms and animals, it can also be produced by chemical synthesis or

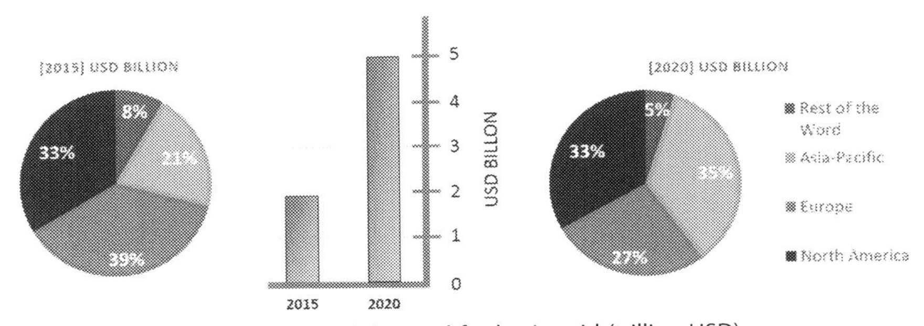

Figure 11.1 Global and regional demand for lactic acid (trillion USD).

by fermentation of carbohydrates. By chemical synthesis, racemic lactic acid is produced from petrochemical resources. And by microbial fermentation of renewable resources, you can obtain a lactic acid L (+) or D (−) optically pure (Wee et al., 2006). And due to application, one form of optically pure lactic acid is preferable over the other.

However, due to the various routes of obtaining it and due to the increasing demand and the various lactic acid markets, there must be improvements in the commercial processes of obtaining said product. The following sections of this chapter will present both the chemical route and the biomass route for obtaining lactic acid. Raw materials, disadvantages and advantages will be raised, and the proposal for the intensification of processes to achieve sustainable processes and frame them within the concept of the circular economy.

11.2 Chemical route of lactic acid production

This route was discovered in 1863 by Wislicenus (Benninga, 1990). It is described by chemical synthesis using the lactonitrile route, which is a byproduct of acrylonitrile technology. Fig. 11.2 describes the chemical synthesis of lactic acid.

11.2.1 Process, raw material, and reactions

Chemical synthesis is carried out with the liquid phase and atmospheric pressure reaction of hydrogen cyanide (HCN) and acetaldehyde (CH_3CHO), catalyzed by a base, to form lactonitrile (C_3H_5NO), which is recovered by distillation and hydrolyzed to lactic acid using HCl or H_2SO_4 concentrate to produce raw lactic acid and ammonium salt. Raw lactic acid is esterified with methanol, producing methyl lactate, then recovered and purified by distillation and hydrolyzed with water under an acid catalyst to produce pure lactic acid and recycle methanol. This product is a colorless liquid, stable to heat. The main disadvantage presented by this method is obtaining racemic mixtures of acid D (−) and L (+) lactic (Gil-Horán et al., 2008).

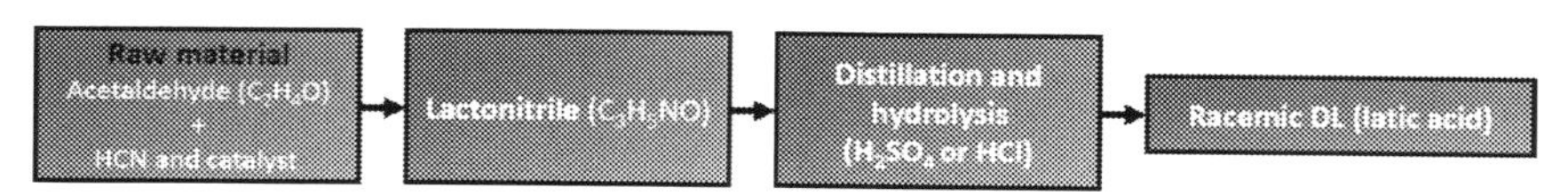

Figure 11.2 Chemical synthesis for the manufacture of lactic acid.

Reactions for the chemical synthesis of lactic acid are described in Eqs. 11.1 − 11.4.

$$CH_3CHO + HCN \rightarrow CH_3CHOHCN \quad (11.1)$$

$$CH_3CHOHCN + 2H_2O + \frac{1}{2}H_2SO_4 \rightarrow CH_3CHOHCOOH + \frac{1}{2}(NH_4)_2SO_4 \quad (11.2)$$

$$CH_3CHOHCOOH + CH_3OH \leftrightarrow CH_3CHOHCOOCH_3 + H_2O \quad (11.3)$$

$$CH_3CHOHCOOCH_3 + H_2O \leftrightarrow CH_3CHOHCOOH + CH_3OH \quad (11.4)$$

To produce one kilogram of lactic acid by chemical synthesis, about 0.515 kg of CH_3CHO, 0.300 kg of HCN, and 0.573 kg of H_2SO_4 are used. This results in a yield of 95% lactic acid (Juodeikiene et al., 2015).

11.2.2 Performance index in lactic acid production via petrochemical

Juodeikiene et al. (2015), show in their work sustainable performance metrics for the chemical synthesis of lactic acid. They quantify energy expenditure for each section of the process, resulting in an energy efficiency of 32%. The total energy input is 52 GJ/ton. The energy input of the most important operations is: 7 GJ/ton in recovery; 6 GJ/ton in distillation, 14 GJ/ton in esterification, and 9 GJ/ton in hydrolysis. In soil use for plantings, compared to the fermentation route, the chemical synthesis of lactic acid does not require available soil to obtain the raw material. However, in the area of the total cost of the process via chemical synthesis, Juodeikiene et al. (2015) estimated a cost of 1225 $USD/ton. Which makes it less competitive in terms of energy and costs than the fermenting route. That is why companies such as *Monsanto*, *Sterling Chemicals*, and *Musashino Chemical* have decided to stop or change their production to fermentation processing.

11.2.3 Disadvantages in the production of lactic acid via petrochemical

There exist various possible routes to produce lactic acid by chemical synthesis. However, the only feasible technical route is the one that uses lactonitrile as a raw material (Gao et al., 2011). The main disadvantages of chemical lactic acid production are that it is costly, and that it depends on by-products from other industries, which are derived from the petrochemical industry (Datta and Henry, 2006). Another important aspect is that the chemical synthesis produces a racemic mixture of lactic acid. However, for numerous specific applications, only one of the isomers of lactic acid is desired (Abdel-Rahman et al., 2011). Problems related to the high cost of raw material, product impurity and dependence on other industries lead to the use of biotechnological processes based on biomass fermentation.

11.3 Conventional process of production of lactic acid via fermentation of biomass

Today the route of obtaining lactic acid is the most widely used (about 90% of total production). Microbial fermentation (see Fig. 11.3) has as its characteristic the biological degradation of a substrate by microorganisms in metabolites. There are numerous raw materials (*Alfalfa fibers*, *Cellobiose*, *Corn Stover*, *Glucose*, etc.) and microorganisms (*Lb. Plantarum*, *E. mundtii QU 25*, *B. coagulants LA204*, *Lb. lactis BME5—18M*, etc.) that can be used in the production of lactic acid (Komesu et al., 2017). Depending on the type of substrate and its purity, high purity lactic acid can be obtained, such as sugarcane sucrose. However, the high cost of sucrose makes it economically not feasible. Despite this, there are several biomasses that make the process profitable and environmentally advantageous.

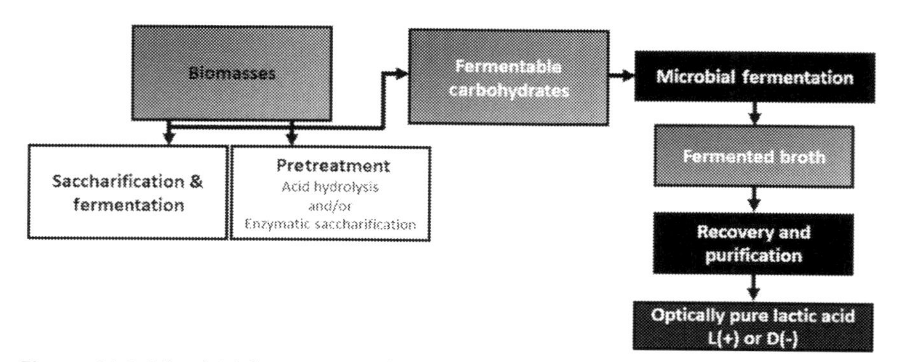

Figure 11.3 Microbial fermentation for the manufacture of lactic acid.

11.3.1 Raw material for the production of lactic acid via biomass

An important aspect in the choice of raw material for microbial fermentation in obtaining lactic acid is cost. There is a large list of raw materials that can be used for lactic acid production, below are some of the specific cases that will be summarized with their pros and cons.

1. An advantage of starched materials (i.e., wheat, corn, potatoes, rice, barley, etc.): is that they can prevent glucose repression. About 90% of commercially available lactic acid is produced by submerged maize fermentation (Wang et al., 2016).
2. Lignocellulose Biomass: can be used to achieve sugar solutions that can be used for the production of lactic acid through the following stages: (1) for retreating, (2) enzymatic hydrolysis and (3) fermentation of sugar to lactic acid (Idler et al., 2015). While the cost of lignocellulose is low, the pretreated stage makes the whole process economically unworkable. Simpler and more promising methods such as hydrothermal pretreatment is currently being developed to make the process more economical and feasibility (Eom et al., 2015).
3. Whey: turns out to be an important and convenient raw material to produce lactic acid (Panesar et al., 2007), by composing nutrients essential for microbial growth (proteins, fats, water-soluble vitamins, mineral salts).
4. Food waste: it has a high potential for lactic acid production since it is usually rich in carbohydrates. In addition, it is advantageous as an effective method of environmental waste management (Tang et al., 2016).
5. Glycerol: the transformation of glycerol into lactic acid can be classified into hydrothermal methods (Kishida et al., 2005) and heterogeneous catalysis methods (Auneau et al., 2012).
6. Microalgae: by not containing lignin, it facilitates conversion into fermentation substrate. Microalgae grow almost anywhere, have an extremely short collection cycle and have high fermentable sugar content (Nguyen et al., 2012).

11.3.2 Lactic acid production via biomass

As can be seen in Fig. 11.3, the production of lactic acid via biomass fermentation consists of a series of continuous processes. The first process is pretreated, this can be carried out by acid hydrolysis and/or by a simultaneous process of Saccharification. Once pretreated, the material is ready to undergo fermentation, depending on the fungal or bacterial strain, production can be used to produce lactic acid D (−) or L (+). Similarly, production strains will govern operating conditions for optimal conversion to lactic acid (Juodeikiene et al., 2015).

11.3.2.1 Fermentation route

Fermentation can generally be described by the following reactions (Juodeikiene et al., 2015):

$$C_6H_{12}O_6 + 2NaOH \rightarrow 2(CH_3CHOHCOO^-)Na^+ + 2H_2O \tag{11.5}$$

$$(CH_3CHOHCOO^-)Na^+ + H_2O \rightarrow CH_3CHOHCOOH + NaOH \tag{11.6}$$

$$CH_3CHOHCOOH + CH_3OH \leftrightarrow CH_3CHOHCOOCH_3 + H_2O \tag{11.7}$$

$$CH_3CHOHCOOCH_3 + H_2O \leftrightarrow CH_3CHOHCOOH + CH_3OH \tag{11.8}$$

Juodeikiene et al. (2015) get an efficiency of 48%. However, when performing an esterification stage, an increase in material efficiency can increase to 68%. However, the fermenting route exhibits some drawbacks. Unwanted by-products, such as calcium sulfate, are produced, among other organic acids, after the fermentation stage, for which the elimination of these by-products is required. Deriving several problems such as the high cost of product recovery and the complex nature of the biological process. The following section shows some of the lactic acid recovery and purification processes.

11.3.2.2 Lactic acid recovery and purification processes

Today in the commercial field there are few cost-effective processes to produce lactic acid. This is mainly due to the high costs associated with the recovery and purification of lactic acid. One aspect to consider is that lactic acid has a strong affinity with water, is a high boiling point component and tends to oligomerize at high temperatures. That is why the work of recovering and purifying lactic acid turns out to be one of the most important stages of the production process.

There are already several processes studied, which aim to reduce cost and make the process more cost effective, here are some of these processes:

1. Membrane separation processes are intermediate phases separating two phases and act as an active or passive barrier in the transport of matter between phases (Rubio et al., 2009).
   a. Ultrafiltration: used to retain relatively large molecules in a range between 2 and 100 nm.
   b. Reverse osmosis: allows retaining molecules of low molecular weight, such as salts. The retention of organic solvents is usually not high because they dissolve in the membrane.
   c. Electrodialysis: membrane separation technology in which ions are transported through an ion exchange membrane from one solution to another, under the influence of an electrical potential.

An important advantage is that high purity can be obtained with the use of these technologies. However, they also have several drawbacks such as high costs (due to high pressures to achieve filtration through the membrane), low performance and a loss of efficiency in operating time.

1. Adsorption: lies in contacting a fluid (liquid or gas) with a solid, the adsorbent. One or more fluid components are brought to the absorbent surface. These components can be separated from others that are less attracted to the surface of the adsorbent (Aljundi et al., 2005).

This technology presents inconveniences of selectivity, capacity and particle regeneration, so it is not inefficient.

1. Solvent extraction: can be defined as the passage from a solute in an aqueous phase to another liquid organic phase, immiscible with it, known as the organic phase, in order to separate it from the other species of the solution (San-Martín et al., 1996).

This technique requires relatively large quantities of solvent, and it is difficult to obtain high purity lactic acid.

1. Precipitation: it is a simple and effective method to separate and recover lactic acid in the form of calcium lactate from fermentation broth by adding sulfuric acid (Min et al., 2011).

It has a low recovery of lactic acid.

Currently, the use of process intensification techniques has shown a visible improvement with the use of reactive distillation to reduce all these limitations shown by the other proposed processes. In this chapter, we aim to show more about the lactic acid recovery process based on reactive distillation, through a modified process using two-way vapor/liquid flows between different columns (thermally coupled). The design and optimization of these intensified processes of recovery and purification of lactic acid will address production from a point of view of sustainability and circular economy.

11.3.3 Advantages and disadvantages of lactic acid production via biomass

The fermentation process for obtaining lactic acid offers several advantages s in terms of the use of renewable carbohydrate biomass, low production temperature, low energy consumption and promotes the production of high selective purity of lactic acid either D (−) or L (+). That is why lately this process has received great interest, as it offers a more environmentally friendly alternative than that provided by the petrochemical industry. A very important advantage over chemical synthesis is that biomass production manages to use cheap raw materials. However, there are still diverse disadvantages that need to be addressed in order to produce lactic acid by fermentation, such as the development of high-performance lactic acid-producing microorganisms and with special attention to achieve low-cost economic and energy recovery and purification processes, as well as for the sustainable and circular economy.

11.3.4 Problems in the production of lactic acid via biomass

The main problem with lactic acid production is the economics of the process. Juodeikiene et al. (2015), show that operating costs in lactic acid production are approximately 77% of total costs. Associating 26% with raw materials, 34% with fermentation, 28% with the recovery process, and 12% with hydrolysis. Similarly, in equipment costs, recovery (electrodialysis and evaporator) for them represents about 40%, being the largest contributor to the capital costs of the process. That is why the use of intensified recovery schemes, such as reactive distillation, and thermally coupled distillation systems, are intended to provide great benefits to achieve an economical and sustainable process.

11.4 Proposals for intensification of the process of obtaining lactic acid via biomass

Process intensification represents an important trend in processing technology and attracts more and more attention in the industry and research community. Reactive distillation and thermally coupled distillation are two promising technologies that achieve substantial economic benefits within process intensification. It has been shown that reactive distillation with thermal coupling can provide better energy efficiency than conventional reactive distillation (Malone and Doherty, 2000). This is because a thermally coupled reactive system has greater energy savings, eliminating the remixing effect present in the conventional reactive system. That is why several strategies were adopted in this chapter to improve the energy performance of reactive distillation systems for lactic acid recovery, which lead to the use of unconventional distillation sequences. It is intended that with the use of thermally reactive distillation columns coupled to the process of producing lactic acid via fermentation, operating and investment costs can be reduced, thus increasing energy efficiency and reducing the number of equipment present.

This chapter shall compare three different types of thermally coupled reactive distillation columns applied to the lactic acid recovery process. This considering a heavy organic impurity and the oligomerization of lactic acid in the process, in order to avoid high energy consumption by remixing. In addition, a thermal coupling study was performed on the hydrolysis column. The study contemplates the synthesis, design, and optimization of purification processes. Optimization was contemplated within economic (measured by Total Annual Cost [TAC]), environmental (measured by [Eco-99]) and inherent security (measured by Individual Risk [IR]) metrics, to obtain sustainable processes and within the concept of the circular economy.

11.4.1 Synthesis and design

Prior to the synthesis and design of reactive distillation columns for the case of the lactic acid production process, it is necessary to establish the feasibility of the technology. To this end, it is verified that this reactive design meets the presence of two or more products. In the same way, it was shown that the operating window of the temperatures between the reaction and the separation is adequate (reaction close to 50°C, and the boiling point for both reagents [CH_3OH and $C_3H_6O_3$] is 64°C – 122°C respectively). And finally, it was established that both pressure and temperature are not close to the critical region of some components. By complying with the above, it is possible to generate the recovery and purification process by means of reactive distillation technology to produce lactic acid (Malone and Doherty, 2000). Once verified above, the design of the intensified configurations was carried out by using the Aspen Plus Simulator V8.8.

The set of reactions carried out within the reactive distillation column is given in Eqs. 11.9 − 11.11.

$$\underbrace{CH_3CH(OH)COOH}_{\text{Lactic Acid}} + \underbrace{CH_3OH}_{\text{Methanol}} \overset{k_{f,1}}{\underset{k_{r,1}}{\longleftrightarrow}} \underbrace{CH_3CH(OH)COOCH_3}_{\text{Metil lactate}} + \underbrace{H_2O}_{\text{water}} \tag{11.9}$$

$$\underbrace{CH_3CH(OH)COOH}_{LacticAcid} + \underbrace{CH_3CH(OH)COOH}_{LacticAcid} \overset{k_{f,2}}{\underset{k_{r,2}}{\longleftrightarrow}} \underbrace{C_6H_{10}O_5}_{Dilactate} + \underbrace{H_2O}_{water} \tag{11.10}$$

$$\underbrace{C_6H_{10}O_5}_{Dilactate} + \underbrace{CH_3CH(OH)COOH}_{Lactic\ \ Acid} \overset{k_{f,3}}{\underset{k_{r,3}}{\longleftrightarrow}} \underbrace{C_9H_{14}O_7}_{Trilactate} + \underbrace{H_2O}_{water} \tag{11.11}$$

Where kinetic equations are presented in a pseudo-homogeneous way. Kinetic data are reported in Table 11.2 (Kim et al., 2017).

To prevent lactic acid from oligomer ($>80°C$), the temperature of the test column dome should be monitored. To avoid oligomerization, attention should be paid to the operating pressure of the column. The design of this column contemplated a catalyst density of 770 kg/m^3, and a 50% tray holdup volume for the solid catalyst. The thermodynamic model used to calculate the equilibrium composition of vapor-liquid and vapor-liquid-liquid, is the *Universal Quasi-Chemical Activity Coefficients* (UNIQUAC). In addition, the *Hayden-O'Connell* equation was used for the steam phase.

A representative feed was used for the recovery and purification scheme with a flow of 50 kmoL/h (8.4% moL of lactic acid, 90.5% moL of water, and 1.1% moL of succinic acid), a feed pressure of 2 atm and a temperature of 35°C. All the designs presented exhibit a column I (preconcentrator), where the fermenter stream is fed and the most water is removed. Once almost all water is removed, and before the azeotropic is added, it is fed to column II. In column II, methanol-assisted esterification (Eq. 11.9) is performed, where pure succinic acid is produced at the bottom of the spine at the same time. The distillate stream is fed to column III, where a hydrolysis

Table 11.2 Kinetic data for reactive distillation.

Reaction	Kinetic model
Methyl lactate	$r = m_{cat}\left(k_f a_{ACL} a_{MEOH} - K_r a_{METILLACTATE} a_{H_2O}\right)$ $k_f = 2.98 \times 10^4 \exp\left(-\frac{51300}{RT}\right)$ $k_r = 2.74 \times 10^2 \exp\left(-\frac{47200}{RT}\right)$
Dilactate	$r = m_{cat}\left(k_f x_{ACL} a_{ACL} - K_r x_{L2} x_{H_2O}\right)$ $k_f = 1.1 \times 10 \exp\left(-\frac{52000}{RT}\right)$ $k_r = 5.54 \times 10 \exp\left(-\frac{52000}{RT}\right)$
Trilactate	$r = m_{cat}\left(k_f x_{L2} a_{ACL} - K_r x_{L3} x_{H_2O}\right)$ $k_f = 4.53 \times 10^{-4} \exp\left(-\frac{50800}{RT}\right)$ $k_r = 2.28 \times 10 \exp\left(-\frac{50800}{RT}\right)$

process is generated to regenerate lactic acid (Eqs. 11.10 and 11.11). And to complete the design, in the last column IV, both the alcohol and water are separated and purified from the esterification, this is to be able to reuse them in the process (it is recirculated to column II).

To carry out the intensification of the conventional sequence, care was taken to thermally couple some of the streams. That is, based on the conventional scheme, condensers or reboilers were eliminated at convenience to generate energy-efficient schemes.

Intensified schemes are presented in Fig. 11.4. For the TC-C and TC-R schemes, both the column III condenser and the column I reboiler were removed, respectively. The Petlyuk scheme was achieved by removing both, condenser and reboiler from column III and thermally attaching it with column IV. And finally, the Petlyuk-R scheme, where apart from the thermal coupling between column III and IV, the reboiler of column I was eliminated and thermally coupled with column II. Thermal coupling between the two columns was avoided with the chemical reaction, to avoid a decrease in reaction performance. All this was designed with the aim of mitigating the energy of the lactic acid purification process, as

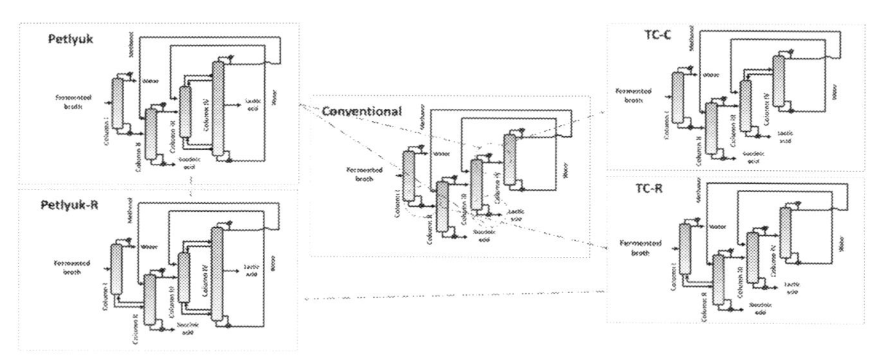

Figure 11.4 Intensified schemes for obtaining lactic acid.

well as generating a circular economy environment, reusing the streams and recirculating them to the process.

11.4.2 Optimization

To optimize the configurations shown above, the stochastic optimization method called differential evolution with Tabu list (DETL) was used. It should be understood that these configurations result in nonlinear and multivariate models, and the objective functions used as optimization criteria are generally nonconvex with several optimal locations. That is why stochastic optimization methods play an important role to solve such processes. The hybrid algorithm is encoded in Visual Basic. Using dynamic data exchange, the numeric method generates input vectors that are evaluated in the process model (Aspen Plus). The model evaluates the input vector and generates output vectors. The hybrid method examines the input and output data and, depending on the evolutionary nature of the algorithm, the process begins to repeat itself. The operational parameters for the stochastic method were: 120 individuals, 900 generations, 50% of the number of individuals as a Tabu list, 0.6 for the crossover, and a Tabu radius of 1×10^{x6}. These parameters were obtained from an adjustment process in previous calculations.

11.4.2.1 Performance indices

The metrics chosen for process optimization were governed by circular economy and sustainability. Both concepts, both the circular economy and sustainability fall within a framework of process economics, environmental character and inherent safety, among others (Geissdoerfer et al., 2017). According to Jiménez-González et al. (2012), in every process it is essential to incorporate "ecological metrics" and this will detonate in process designs framed in the circular economy. This chapter aims to analyze the different topologies, based on a conventional design modification, for the reactive distillation process for the purification of lactic acid and select the most promising according to various indices.

11.4.2.1.1 Economic index

TAC was used to assess the economic performance of the process. To this end, the methodology set out by Turton et al. (2008) was followed. Eq. 11.12 shows the calculation of the TAC.

$$TAC = \frac{\sum_{i=1}^{n} C_{TM,i}}{r} + \sum_{j=1}^{n} C_{ut,j} \tag{11.12}$$

where C_{TM} is the total cost of the module, symbolizes C_{BM}^{0} the cost of the bare module, expresses the direct and indirect costs of each unit, r the number of years of return on investment, n is the total number of individual units, is the cost of services, represents the cost of the $F_{BM}^{0} C_{TM,i} C_{ut,j} C_{BM,i}^{0}$ bare module. For this chapter, five years were considered as a recovery period. Assuming the plant works 8500 h a year. And with the associated expenses of: pressure steam (HP) (42 bar, 254°C, \$9.88 G/J), medium-press steam (MP) (11 bar, 184°C, \$8.22 G/J), low-pressure steam (LP) (6 bar, 160°C, \$7.78 G/J) and cooling water (\$0.72 G/J) (Luyben, 2012).

11.4.2.1.2 Environmental index

Eco-indicator 99 (EI99) was chosen as the quantitative indicator of environmental damage. This indicator is based on the life cycle assessment and the approach was proposed by Goedkoop (2007). The EI99 is calculated as follows:

$$\text{EI99} = \sum_{b} \sum_{d} \sum_{\text{keK}} \delta_d \omega_d \beta_b \alpha_{b,k} \tag{11.13}$$

Where β_b represents the total amount of chemical b released per unit of reference flow due to direct emissions, α_b, k_{is} the damage caused in category k per unit of chemical b released into the environment, d_{is} a weighting factor for damage in category d, and δ_d is the normalization factor for category d damage. This is with respect to three main categories of impact (human health, ecosystem quality, and resource depletion).

11.4.2.1.3 Inherent safety index

Inherent safety was quantified by the individual risk (IR) index. IR is defined as the risk of injury or death of a person in the vicinity of a hazard (Freeman, 1990). The primary objective of this index is to estimate the probability of affectation caused by the precise incident occurring at a certain frequency. The IR does not depend on the number of people exposed. Eq. 11.14 shows the calculation of IR.

$$\mathrm{IR} = \sum f_i P_{x,y} \tag{11.14}$$

Where f_i is the frequency of occurrence of incident i, while $P_{x,y}$ is the probability of injury or death caused by incident i.

11.4.2.2 Optimization results

It should be noted that purity limits associated with commercial requirements were established (90% by weight for lactic acid, 99% by weight for succinic acid, 99.9% by weight of methanol, and water to recycle). Figs. 11.5–11.7 show the results of the optimization process (108,000 evaluations were conducted). For greater understanding, the results are presented in two-dimensional charts.

At first glance, it can be seen in the results of Figs. 11.5–11.7, that the lowest TAC and EI99 values are associated with the Petlyuk-R sequence.

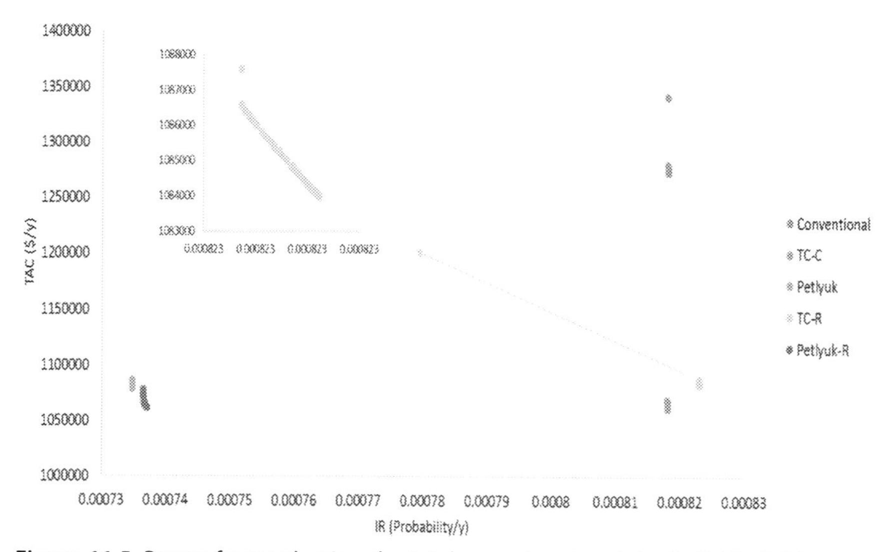

Figure 11.5 Pareto front valuating the total annual cost and the individual risk.

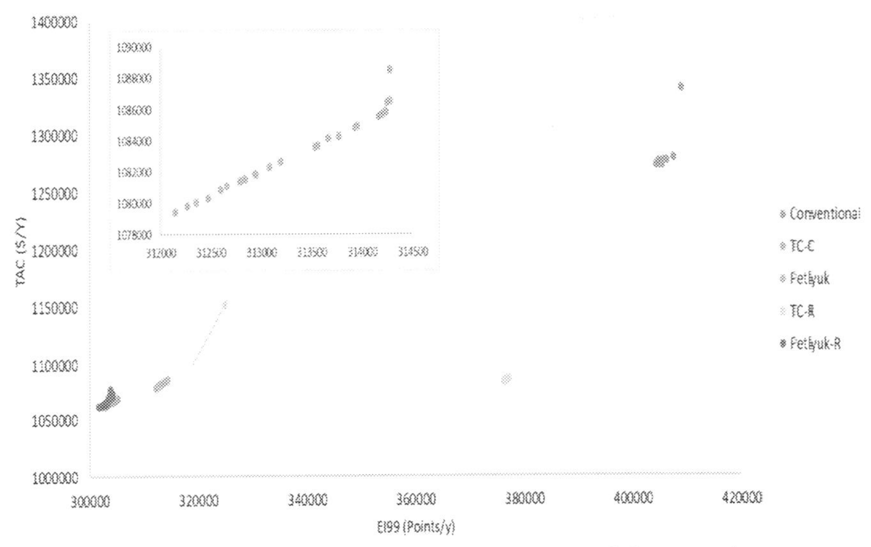

Figure 11.6 Pareto front valuating the total annual cost and the eco-indicator 99.

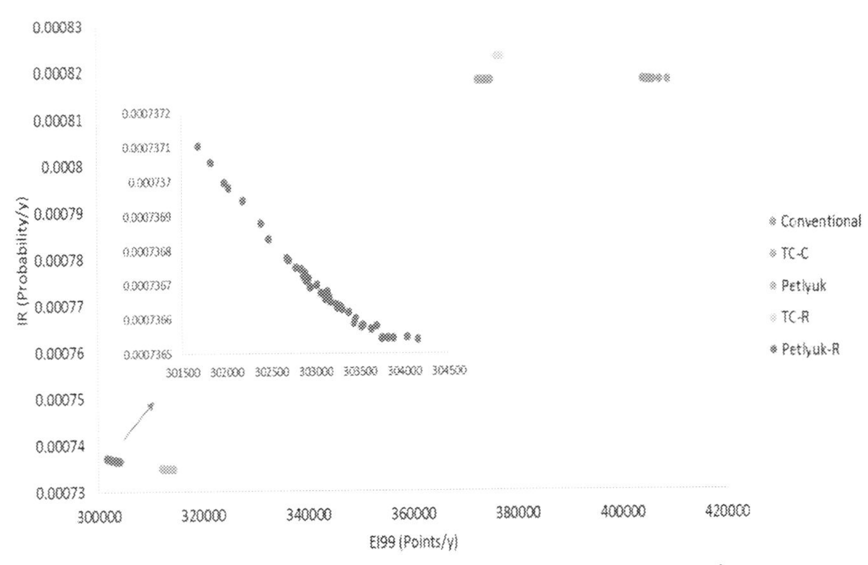

Figure 11.7 Pareto front evaluating the individual risk and the eco-indicator 99.

In addition, to present itself as the second-best sequence in terms of IR. Despite this, an in-depth analysis of each of the sequences must be carried out to see potential advantages and examine which of them is most convenient in terms of the circular economy and sustainability.

It is evident that the trend of improvement of each of the indicators (TAC, EI99, and IR) is the more intensified sequence, in terms of reboilers. However, the same is not true when removing condensers. The values of TC sequences with respect to the conventional scheme in terms of TAC are given as follows:

1. Sequence TC-R 11% TAC reduction in column I bottom coupling with column II. However, the reboiler duty in column II increased by 76%, this is because this expenditure is carried out in the reboiler of that column.
2. Sequence TC-C reduction of 17%. These improvements are associated with energy savings by replacing external services with interconnection flows. This sequence showed a 14% reduction in reboiler duty.

With respect to environmental impact, both sequences have similar behavior. The values associated with the IR show that the more exchange of flows at high temperatures exist between the equipment, slightly increased the risks associated with their operation. That's why the TC-R sequence has a 1% increase in IR over the conventional scheme.

The schemes with the highest degree of intensification present an interesting phenomenon. For the Petlyuk sequence, there is an increase in TAC, despite energy savings when using interconnection flows. For EI99 and IR indicators, they exhibit opposite behaviors, there is a slight increase in environmental load (3% over TC-C) and a considerable drop in danger when IR is reduced by 10%. The Petlyuk-R scheme has very similar values in TAC and EI99 with respect to Petlyuk, and a very slight increase in IR.

With the above, it can be concluded that in lactic acid recovery and purification sequences, thermal couplings in the reboiler area do not improve their overall performance. Otherwise when the coupling is presented in the condenser. If we consider the circular economy and sustainability as motivation, it is indisputable that a thermal coupling is better. As an example, TC sequences are noted to significantly improve on the circular economy metrics evaluated. An important point to note is that hydrolysis and esterification reactions appear to be disadvantaged by the proposed interconnection flows. However, by using the optimization tool, it helps to know the right area and flows to maintain the performance and productivity of the process.

Tables 11.3 and 11.4 exhibit the main design parameters for all proposed sequences. It is indisputable that there are significant changes in the resulting designs, and this is due to the energy disturbances achieved by the proposed couplings. Therefore, the lower energy consumption there

Table 11.3 Design parameters of the conventional, TC-C, and TC-R sequences.

	Conventional				TC-C				TC-R			
Columns	CI	CII	CIII	CIV	CI	CII	CIII	CIV	CI	CII	CIII	CIV
Stages	43	8	60	32	42	8	72	31	46	8	60	31
Feed stage	23	4,4	42,42	14	24	5,4	41,46	17	15	4,4	40,34	23
Side stage								16	46			
Reflux ratio	0.1	0.05	0.12	1.82	0.04	0.026	-	1.77		0.29	0.05	0.9
Distillate flow (kmoL/h)	40.4153	25.734	34.673	16.64	40.41	25.73		16.73	40.4	25.7	34.7	16.6
Bottoms flow (kmoL/h)	9.58	0.53	5.05	18.03	9.58	0.53	5.05	18.03	-	0.53	5.05	9.58
Interconnection flow (kmoL/h)							5.899		14.5858			
Diameter (m)	1.03	1	0.86	1.02	1.03	1	0.86	1.02	1.03	1	0.86	1.02
Reactive stages		3 to 7	28 to 47			3 to 7	27 to 66			2 to 7	27 to 48	
Holdup (L)		29.14	97.062			28.9034124	91.6844112			30.5	96.6	
Pressure (atm)	0.1902	0.4599			0.211	0.37			0.12	0.37		
Reboiler duty (cal/s)	448115				338868				417337			
Succinic acid (wt.%)		0.999				0.999				0.99		
Lactic acid (wt.%)			0.94				0.914				0.95	
Methanol (wt.%)				0.999				0.995				0.99
TAC ($/year)	1,274,559				1,062,454				1,134,718			
EI99 (eco-points/year)	405,025				302,771				376,834			
IR (probability/year)	0.000818084				0.000818023				0.000823006			

Table 11.4 Design parameters of the Petlyuk, and Petlyuk-R sequences.

	Petlyuk				Petlyuk-R			
Columns	C1	C2	C3	C4	C1	C2	C3	C4
Stages	43	9	76	26	44	9	60	23
Feed stage	20	5,7	28,62	7,25	15	4,4	40,34	23
Vapor interconnection flow (kg)			36.75		14.5858			
Input stage (vapor)			76		46			
Side stream (vapor)				25		5, 25		21
Side stream (liquido)				6		6		6
Side stream (water)				21		21		17
Reflux ratio	0.1	0.05		1.8518	0.1	0.29	0.05	0.9
Distillate flow (kmoL/h)	40.4	25.7		16.65	40.4	25.7	34.7	16.6
Bottoms flow (kmoL/h)	9.58	0.53	–	5.04	–	0.53	–	5.08
Diámetro (m)	1.03	1	0.86	1.02	1.03	1	0.86	1.02
Reactive stages		4 to 7	26 to 66			2 to 7	27 to 44	
Holdup (L)		42.37	90.07			30.5	96.6	
Pressure (atm)	0.194	0.569	1	1	0.12	0.37	1	1
Reboiler duty (cal/s)	349927				338292			
Succinic acid (wt.%)		0.999				0.99		
Lactic acid (wt.%)			0.958				0.95	
Methanol (wt.%)				0.999				0.99
TAC ($/year)	1086548				1062449			
EI99 (eco-points/year)	313189				303062			
IR (probability/year)	0.000734825				0.000736675			

Figure 11.8 Temperature and composition profiles for all sequences.

is, it can move design variables, reflux ratio, or stages, for example, to minimize objective functions.

Fig. 11.5 shows the antagonistic relationship between TAC and IR. That is, the lower the risk of accidents, the CT is increased. This is largely due to the operation variables and the nature of the mixture to be separated. Low risk of the process is observed when the reflux ratio is raised. Resulting, in a direct increase the energy requirement and therefore of the TAC. In Fig. 11.6, the opposite behavior is displayed between TAC and EI99. This is because in the evaluation of EI99 are considered: steam (this has greater impact), steel and process electricity. That is why the optimization tool tries to find designs that achieve balance in these areas. However, if they move away from stable design parameters, the TAC tends to increase. And finally, Fig. 11.7, exhibits the trend between EI99 and IR. This is the same case in Fig. 11.6, if you want to reduce the environmental impact, you must reduce the energy requirement. And this is achieved with designs where the reflux ratio is relatively low. This is associated with the higher the flow returned to the column, the higher the amount of water, which generates low concentrations of the organic compound. Another point to note is that the separate water in column I, and the recovered methanol, are recycled to the hydrolysis and esterification column, respectively. This creates an atmosphere of circular economy and sustainability in the processes presented in this chapter. It is proposed that the intensified designs presented are very attractive if applied on an industrial level. Fig. 11.8A and B expose the composition and temperature profiles of the sequence presented.

11.5 Conclusions

In this chapter, several alternatives are presented for the separation and purification of lactic acid. These alternatives were optimized and evaluated under three objective functions in a circular economy framework (TAC, EI99, and IR). Having optimized designs, it was stated that replacing the reboilers with interconnection flows has no significant impact on economic or environmental cost, rather it poses a danger in terms of process safety. Otherwise, by replacing the condensers with interconnecting flows, all indices improved. Petlyuk schemes were the ones that potentially saved on all three indicators (17% in TAC, 10% in EI99, and 10% in IR). By looking at the results of the optimal sequences, we can relate the interaction of the design variables to each of the indexes. That is, the higher the reflux ratio, it directly impacts energy consumption and generates an increase in indices. Finally, the sequence with the best results is the one that presents the thermal coupling between the first two columns followed by a Petlyuk column, the above based on a circular economy framework. The Petlyuk-R sequence performs with characteristics based on the circular economy and the results may indicate the possibility of using these industrially intensified technologies for the purification of lactic acid with good indicators of circular economy.

References

Abdel-Rahman, M.A., Tashiro, Y., Sonomoto, K., 2011. Lactic acid production from lignocellulose-derived sugars using lactic acid bacteria: overview and limits. J. Biotechnol. 156 (4), 286–301.

Abdel-Rahman, M.A., Tashiro, Y., Sonomoto, K., 2013. Recent advances in lactic acid production by microbial fermentation processes. Biotechnol. Adv. 31 (6), 877–902.

Aljundi, I.H., Belovich, J.M., Talu, O., 2005. Adsorption of lactic acid from fermentation broth and aqueous solutions on Zeolite molecular sieves. Chem. Eng. Sci. 60 (18), 5004–5009.

Altıok, D., 2004. *Kinetic modelling of lactic acid production from whey*. Master's thesis, Izmir Institute of Technology.

Auneau, F., Arani, L.S., Besson, M., Djakovitch, L., Michel, C., Delbecq, F., et al., 2012. Heterogeneous transformation of glycerol to lactic acid. Top. Catal. 55 (7–10), 474–479.

Benninga, H., 1990. A History of Lactic Acid Making: A Chapter in the History of Biotechnology, vol. 11. Springer Science & Business Media.

Chang, D.E., Jung, H.C., Rhee, J.S., Pan, J.G., 1999. Homofermentative production of d-or L-lactate in metabolically engineered Escherichia coli RR1. Appl. Environ. Microbiol. 65 (4), 1384–1389.

Corma, A., Iborra, S., Velty, A., 2007. Chemical routes for the transformation of biomass into chemicals. Chem. Rev. 107 (6), 2411–2502.

Danner, H., Madzingaidzo, L., Thomasser, C., Neureiter, M., Braun, R., 2002. Thermophilic production of lactic acid using integrated membrane bioreactor systems coupled with monopolar electrodialysis. Appl. Microbiol. Biotechnol. 59 (2–3), 160–169.

Datta, R., Henry, M., 2006. Lactic acid: recent advances in products, processes and technologies—a review. J. Chem. Technol. Biotechnol. 81 (7), 1119–1129.

Eom, I.Y., Oh, Y.H., Park, S.J., Lee, S.H., Yu, J.H., 2015. Fermentative l-lactic acid production from pretreated whole slurry of oil palm trunk treated by hydrothermolysis and subsequent enzymatic hydrolysis. Bioresour. Technol. 185, 143–149.

Freeman, R.A., 1990. CCPS guidelines for chemical process quantitative risk analysis. Plant/Operations Prog. 9 (4), 231–235.

Gao, C., Ma, C., Xu, P., 2011. Biotechnological routes based on lactic acid production from biomass. Biotechnol. Adv. 29 (6), 930–939.

Geissdoerfer, M., Savaget, P., Bocken, N.M., Hultink, E.J., 2017. The circular economy–a new sustainability paradigm? J. Clean. Prod. 143, 757–768.

Gil-Horán, R.H., Domínguez-Espinosa, R.M., Pacho-Carrillo, J.D., 2008. Bioproducción de ácido láctico a partir de residuos de cáscara de naranja: procesos de separación y purificación. Tecnol. Ciencia, Educ. 23 (2), 79–90.

Goedkoop, M., 2007. The eco-indicator 99 methodology. J. Life Cycle Assess. Jpn. 3 (1), 32–38.

Idler, C., Venus, J., Kamm, B., 2015. Microorganisms for the production of lactic acid and organic lactates. Microorganisms in Biorefineries. Springer, Berlin, Heidelberg, pp. 225–273.

Jiménez-González, C., Constable, D.J., Ponder, C.S., 2012. Evaluating the "Greenness" of chemical processes and products in the pharmaceutical industry—a green metrics primer. Chem. Soc. Rev. 41 (4), 1485–1498.

Juodeikiene, G., Vidmantiene, D., Basinskiene, L., Cernauskas, D., Bartkiene, E., Cizeikiene, D., 2015. Green metrics for sustainability of biobased lactic acid from starchy biomass versus chemical synthesis. Catal. Today 239, 11–16.

Keller, R., 2007. *Biofuel from Algae has problems, Ag Professional*, June 11, 2011.

Kim, S.Y., Kim, D.M., Lee, B., 2017. Process simulation for the recovery of lactic acid using thermally coupled distillation columns to mitigate the remixing effect. Korean J. Chem. Eng. 34 (5), 1310–1318.

Kishida, H., Jin, F., Zhou, Z., Moriya, T., Enomoto, H., 2005. Conversion of glycerin into lactic acid by alkaline hydrothermal reaction. Chem. Lett. 34 (11), 1560–1561.

Komesu, A., de Oliveira, J.A.R., da Silva Martins, L.H., Maciel, M.R.W., Maciel Filho, R., 2017. Lactic acid production to purification: a review. BioResources 12 (2), 4364–4383.

Long, N.V.D., Lee, M., 2017. Advances in Distillation Retrofit. Springer.

Luyben, W.L., 2012. Principles and Case Studies of Simultaneous Design. John Wiley & Sons.

Malone, M.F., Doherty, M.F., 2000. Reactive distillation. Ind. Eng. Chem. Res. 39 (11), 3953–3957.

Min, D.J., Choi, K.H., Chang, Y.K., Kim, J.H., 2011. Effect of operating parameters on precipitation for recovery of lactic acid from calcium lactate fermentation broth. Korean J. Chem. Eng. 28 (10), 1969.

Nguyen, C.M., Kim, J.S., Hwang, H.J., Park, M.S., Choi, G.J., Choi, Y.H., et al., 2012. Production of L-lactic acid from a green microalga, Hydrodictyon reticulum, by Lactobacillus paracasei LA104 isolated from the traditional Korean food, makgeolli. Bioresour. Technol. 110, 552–559.

Panesar, P.S., Kennedy, J.F., Gandhi, D.N., Bunko, K., 2007. Bioutilisation of whey for lactic acid production. Food Chem. 105 (1), 1–14.

Park, E.Y., Anh, P.N., Okuda, N., 2003. Bioconversion of waste office paper to L (+)-lactic acid by filamentous fungus *Rhizopus oryzae*. Macro Rev. 16 (1), 281–285.

Rubio, O.A.P., Jørgensen, S.B., Jonsson, G.E., 2009. Lactic acid recovery in electro-enhanced dialysis: modelling and validation, Computer Aided Chemical Engineering, vol. 26. Elsevier, pp. 773–778.

San-Martín, M., Pazos, C., Coca, J., 1996. Liquid–liquid extraction of lactic acid with alamine 336. J. Chem. Technol. Biotechnol. 65 (3), 281–285.

Serna-Cock, L., Stouvenel, A.R.D., 2005. Produccion biotecnologica de acido lactico: estado del arte biotechnological production of lactic acid: state of the art produccion biotecnoloxica de acido lactico: estado do arte. CYTA-J. Food 5 (1), 54–65.

Suriderp, C. (1995). Ullman's Encyclopedia of Industrial Chemistry: ácido láctico, fifth ed., 97–104.

Tang, J., Wang, X., Hu, Y., Zhang, Y., Li, Y., 2016. Lactic acid fermentation from food waste with indigenous microbiota: effects of pH, temperature and high OLR. Waste Manag. 52, 278–285.

Turton, R., Bailie, R.C., Whiting, W.B., Shaeiwitz, J.A., 2008. Analysis, Synthesis and Design of Chemical Processes. Pearson Education.

Vick Roy, T.B., 1985. Lactic acid. Compr. Biotechnol. 3, 761–766.

Wang, Y., Meng, H., Cai, D., Wang, B., Qin, P., Wang, Z., et al., 2016. Improvement of L-lactic acid productivity from sweet sorghum juice by repeated batch fermentation coupled with membrane separation. Bioresour. Technol. 211, 291–297.

Wee, Y.J., Kim, J.N., Ryu, H.W., 2006. Biotechnological production of lactic acid and its recent applications. Food Technol. Biotechnol. 44 (2), 163–172.

CHAPTER 12

Future insights in bio-based chemical building blocks

Contents

12.1 Future insights in bio-based chemical building blocks

Chemicals are an essential part of many products and services that modern society makes use of. Chemicals legislation needs to remain fit to guarantee the continued high level of protection of health and the environment. The chemical industry is the world's second largest manufacturing industry and has an important role in everything from daily products to innovative technologies. The Organisation for Economic Co-operation and Development (OECD) predicts that the production, use, and trade of chemicals will more than double between 2020 and 2050. However, still two-thirds of chemicals produced in the EU are hazardous to human health and one-third is hazardous to the environment. Therefore, inaction linked to chemical exposures has substantial costs for society (Patel et al., 2006).

The production, use, and trade of chemicals is a core part of the industry and thus not only affects human health and the environment through exposure to hazardous chemicals but also directly affects climate, biodiversity, ecosystems, resilience to climate and environmental risks, the realization of the circular economy, and the use of resources. Therefore, chemical management is an integral part of environmental sustainability. Thus, it should be considered in the development of both green deal initiatives and the chemicals policies how these considerations contribute to:

1. fulfilling sustainable development goals
2. safe chemicals and progressive substitution
3. climate neutrality

Improvements in Bio-Based Building Blocks Production Through Process Intensification and Sustainability Concepts
DOI: https://doi.org/10.1016/B978-0-323-89870-6.00012-3

4 circularity
5 use of renewable resources.

Some chemicals strategy for sustainability will be able to contribute with important elements that are also crucial for the implementation of the circular economy actions, including a sustainable products policy, and other chemicals related strategies (Hermann et al., 2007).

The bio-based economy is an economy in which crops and waste streams from agriculture and the food industry, the so-called renewable resources or biomass, are used for nonfood applications such as materials, chemicals, transport fuels, and energy. The chemical sector is currently using more and more biomass as raw material to replace increasingly scarce mineral oil. Green building blocks are experiencing enormous growth as a result of government funding for sustainable production and CO_2 reductions, as well as consumer demands for environmentally friendly products and their willingness to pay for them. The so-called drop-ins, building blocks that are chemically identical to their petrochemical counterparts, are especially booming. A major benefit of the new generation of bio-based chemicals based on green building blocks is that they are able to compete with petrochemical products in terms of performance; the characteristics are sometimes identical and occasionally even better. As the price of crude raw materials remains one of the main factors in the economic feasibility of bio-based products, capacity increases on locations where low-cost biomass is available. The transition from petrochemical raw materials to renewable raw materials is a step-by-step process. Companies initially work with raw materials that are as pure as possible, like sugar from sugar cane, and starch from wheat or maize. If this is successful they can take the next step; for example, sugar from lignocellulosic biomass. These sugar streams, however, are often less pure which makes producing chemical building blocks far more challenging. Additionally, using lignocellulosic biomass as source for sugars results in residues such as lignin, for which an application to ensure an economically feasible process has yet to be found (Hermann and Patel, 2007).

Chemical building blocks that are identical to their petrochemical counterparts can immediately be integrated into the current industrial infrastructure and make a material completely or partially bio-based. Examples include n-butanol and adipic acid. These building blocks are currently being produced in large volumes from petrochemical raw materials and can be replaced by their bio-based counterparts. There are also examples of new chemicals and materials from renewable raw materials with unique characteristics that are difficult or impossible to produce from petrochemical raw materials. Examples include lactic acid, 1,3-Propanediol (PDO), isobutanol, succinic acid, and furans. New markets are developing around these building blocks. Especially building blocks that can be used in many different polymer groups due to their chemical structure are especially promising and will undergo substantial growth (BACAS, 2004).

What will we expect in the coming years in the area of bio-based chemical building blocks? In order to respond this question, Fig. 12.1 shows the commercial status of 25 of the most important biochemicals reported by Taylor et al. (2015). This status is based on the Technology Readiness Level, an index operates many factors to determine the industrial statutes such as current production capacity, number of facilities, and their type and locations of facilities among others. Based on Fig. 6.1, it is expected that in the following years the efforts will be focused on improvements or future commercial implementations of many of those products.

Green and sustainable chemistry will play a pivotal role in this century to produce feedstock chemicals and fuels derived from biomass. White biotechnology, for the conversion of fermentable sugars to chemicals, is now at an advanced stage. Efficient breakdown and conversion of lignocellulosic material to chemicals and fuels remains one of the biggest

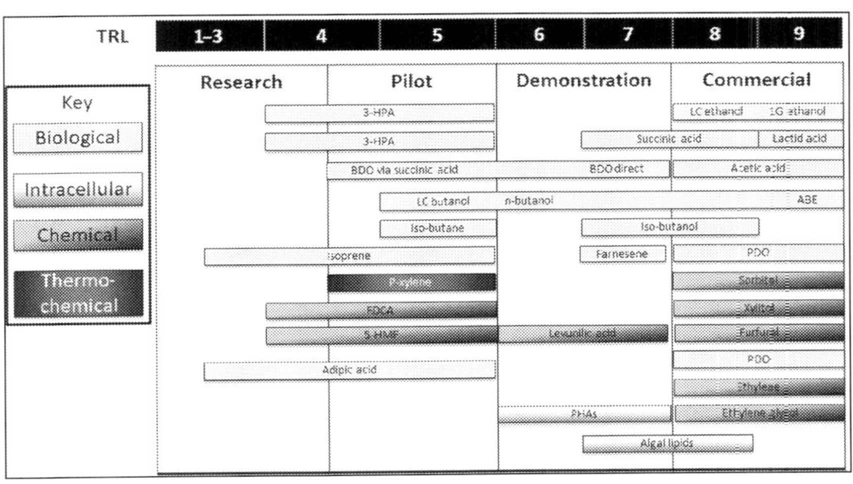

Figure 12.1 Current industrial status of some of the most important biochemicals. *3-HPA*, 3-Hydroxypropanoic Acid; *BDO*, 1,4-Butanediol; *FDCA*, 2,5-Furandicarboxylic acid; *HMF*, Hydroxymethylfurfural; *ABE*, acetone/butanol/ethanol; *PDO*, 1,3-propanediol; *PHA*, Polihidroxialcanoatos; *LC*, Liquid chromatography.

obstacles currently holding back the development of successful biomass-based biorefineries that can compete with traditional petroleum refineries. To improve the modeling of experimental data in bioprocesses for kinetics and equilibrium data. Overcoming lignocellulose recalcitrance to release the locked polysaccharides is important. Several lignocellulose pretreatment technologies are under intensive investigation at the laboratory scale and in pilot plants, for example, dilute or concentrated acid, alkali, flow-through, ammonia fiber explosion, ammonia recycle percolation, lime, steam explosion, and organosolv pretreatment. Development of a lignocellulose pretreatment featuring modest reaction conditions is highly desired; one approach is hydrothermal hydrolysis and/or liquid hot water hydrolysis. Super and subcritical solvents are also been explored. With the advances made in materials science, it should be possible today to prepare more selective and "realistic" catalysts that give the necessary improvements in reactivity. Moreover, these advances should allow for the production of a more environmentally friendly process than those previously reported and even used today. Meanwhile, new materials will open new, more efficient synthetic routes to achieve the final products. The possibility of producing catalysts with well-defined multiple sites that can also include transition metal complexes and enzymes in collaboration with solid acid—base or redox catalysts should allow several reaction steps to be performed in a cascade, avoiding costly intermediate separations which have a strong negative economic impact on the process. The development of relatively small-scale biorefinery concepts (for bioblocks production) that make use of local or regional biomass/waste resources seems to be the most favored method to introduce more advanced (green, whole-crop, and lignocellulosic) biorefinery processes into the market in the medium term (application of concepts on supply chains, process scheduling, and location of biochemical plants). These concepts will require a reduced initial investment, which potentially can be an advantage for the industrial stakeholder in the introduction of new risky initiatives. Most importantly, these initiatives will change the perception of our current society, creating a new socioeconomic and environmental perspective for the future introduction of more advanced biorefinery concepts (e.g., lignocellulosic and marine feedstocks) at a larger scale in the long term. An integrated approach using best available technologies within the infrastructure of a biorefinery needs to be adopted. Introduction of the concepts of circular economy and the water-energy-food nexus in the production of bioblocks. In summary, the four core requirements that must be fulfilled for a biorefinery to be successful: (1) substantial technological breakthroughs must be realized in the bioprocess step, (2) major progress must be made in downstream processing, (3) prices for fossil fuels must be high, and (4) prices for fermentable sugar must be low (EuropaBio, 2003).

Finally, the bioeconomy has become a leading initiative in many regions encouraged by legislation, incentives, and research. The emergence of the circular economy is very consistent with this especially when biorefinery feedstocks are wastes and products are fully recyclable or biodegradable. Certainly, experience with bio-based chemical blocks has taught to be very careful over the choice of feedstocks and that true sustainability goes far beyond using renewable resources. An increasing emphasis on waste feeds will also help to solve the increasing problem of pollution. We must also ensure that the technologies of future biorefineries are themselves highly efficient and of low environmental footprint. The traditional chemical industry was based on nonrenewable feedstocks and often inefficient, wasteful, and dangerous processes. The biorefinery offers us a unique opportunity to make a major step change and provide future societies with genuinely green and sustainable bio-based chemical blocks.

References

BACAS, 2004. Industrial Biotechnology and Sustainable Chemistry. Royal Belgian Academy Council of Applied Science, Brussel.

EuropaBio, 2003. White Biotechnology: Gateway to a More Sustainable Future. European Association for Bioindustries, Brussels.

Hermann, B.G., Blok, K., Patel, M., 2007. Producing bio-based bulk chemicals using industrial biotechnology saves energy and combats climate change. Environ. Sci. Technol. 41, 7915.

Hermann, B.G., Patel, M., 2007. Today's and tomorrow's bio-based bulk chemicals from white biotechnology—a techno-economic analysis. Appl. Biochem. Biotechnol. 136, 361.

Patel, M., Crank, M., Dornburg, V., Hermann, B., Roes, L., Hüsing, B., et al., 2006. Medium and Long-term Opportunities and Risks of the Biotechnological Production of Bulk Chemicals From Renewable Resources—The BREW Project. Utrecht University, Utrecht.

Taylor, R., Nattrass, L., Alberts, G., Robson, P., Chudziak, C., Bauen, A. et al. 2015. *From the sugar platform to biofuels and biochemicals: final report for the European Commission Directorate–General Energy*. E4tech/Re-CORD, Wageningen UR.

Index

Note: Page numbers followed by "*f*" and "*t*" refer to figures and tables, respectively.

C

Printed in the United States
by Baker & Taylor Publisher Services